Eggs and Egg Farms

The Successful Production of Eggs and the Construction of Poultry Houses

by Reliable Poultry Journal

with an introduction by Jackson Chambers

This work contains material that was originally published in 1907.

This publication is within the Public Domain.

This edition is reprinted for educational purposes and in accordance with all applicable Federal Laws.

Introduction Copyright 2017 by Jackson Chambers

COVER CREDITS

Front Cover
Eggs and Egg Farms (Original Cover)
by *Reliable Poultry Journal Publishing Co.*
Photo-manipulation and Restoration by *Self Reliance Books*

Back Cover
Girl With a Basket of Eggs by Joachim Beuckelaer
[Public domain], via Wikimedia Commons

Research / Sources
Wikimedia Commons
www.Commons.Wikimedia.org

Many thanks to all the incredible photographers, artists,
researchers, and archivists who share their great work.

PLEASE NOTE :
As with all reprinted books of this age that are intended to perfectly reproduce the original edition, considerable pains and effort had to be undertaken to correct fading and sometimes outright damage to existing proofs of this title. At times, this task can be quite monumental, requiring an almost total rebuilding of some pages from digital proofs of multiple copies. Despite this, imperfections still sometimes exist in the final proof and may detract slightly from the visual appearance of the text.

DISCLAIMER :
Due to the age of this book, some methods or practices may have been deemed unsafe or unacceptable in the interim years. In utilizing the information herein, you do so at your own risk. We republish antiquarian books without judgment or revisionism, solely for their historical and cultural importance, and for educational purposes.

Self Reliance Books

Get more historic titles on animal and stock breeding, gardening and old fashioned skills by visiting us at:

http://selfreliancebooks.blogspot.com/

INTRODUCTION

I am so pleased to present to you another wonderful old book on poultry production – *Eggs and Egg Farms : The Successful Production of Eggs and the Construction Plans of Poultry Houses.* It was was written by the *Reliable Poultry Journal Publishing Co.*, and first published in 1907, making it more than a century old.

This book focuses on egg production and covers all aspects of running a successful egg production operation, with topics like *Breeding for Egg Production, Egg Yielding Capacity of Hens, Breeding for Eggs, Preservation of Eggs in Cold Storage, Successful Egg Production, Winter Egg Production,* and also discusses various chicken breeds and the different types of poultry houses.

This old publication is a great read for anybody considering embarking on egg production on any scale, and an essential addition to the library of all those who have already decided to take the plunge.

(Note : there are no actual physical "plans" for poultry houses, contrary to the title's suggestion, but there *is* much discussion on poultry houses.)

Jackson Chambers

State of Jefferson, November 2017

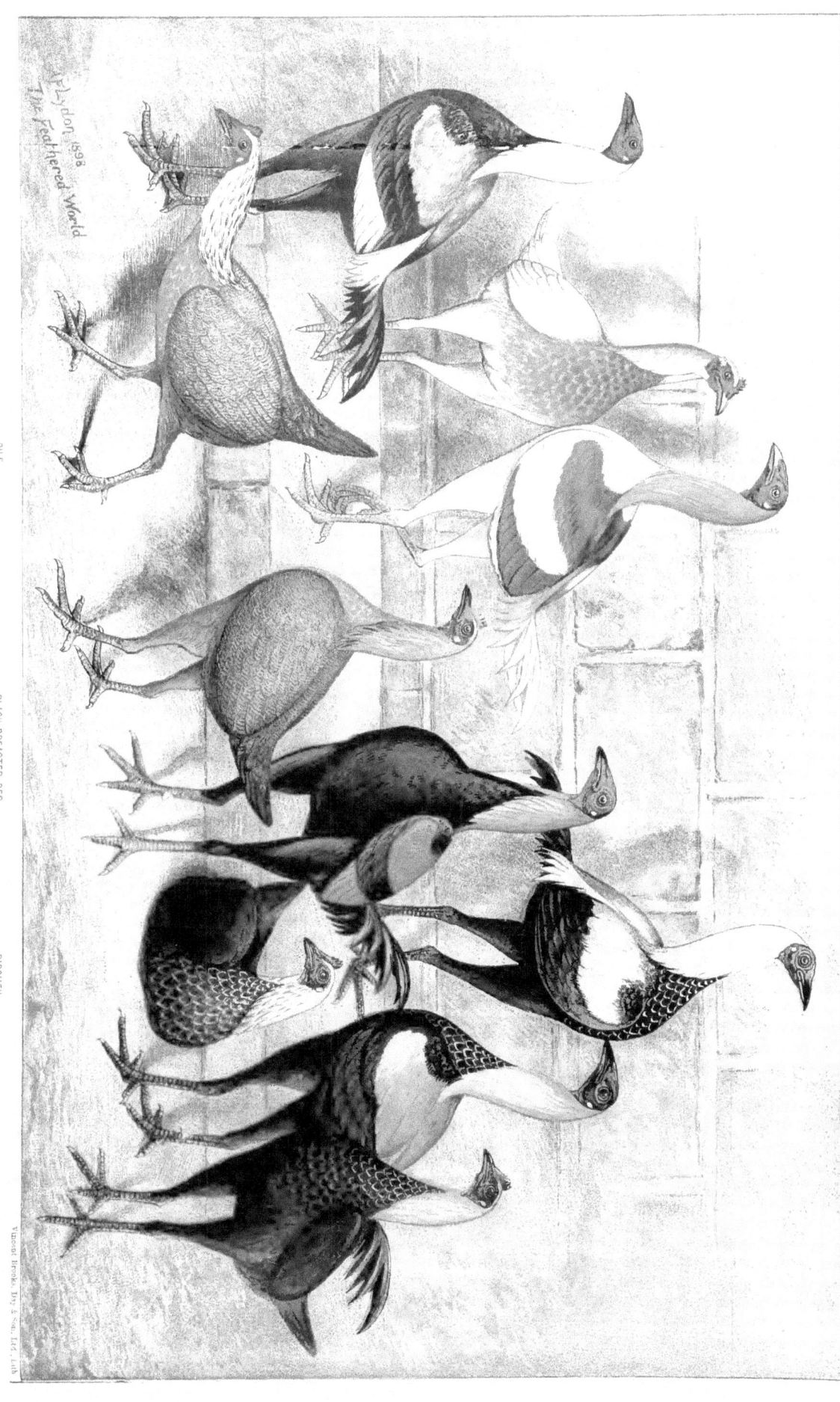

SUPPLEMENT TO The Feathered World.

DUCKWING. PILE. BLACK-BREASTED RED. BIRCHEN. BROWN-BREASTED RED.

GAME BANTAMS.

(Specially drawn to illustrate Mr. Proud's articles on Bantams)

[Supplement to The Feathered World, July 3rd, 1896.]

BLACK ROSECOMB.
JAPANESE.

SILVER AND GOLDEN SEBRIGHTS.
BANTAMS.
BRAHMAS.

WHITE ROSECOMB.
BOOTED.

A TRIO OF SELECTED LAYERS

INTRODUCTION

WE have endeavored to obtain for the readers of *Eggs and Egg Farms* the latest trustworthy information in reference to this important and profitable branch of the poultry industry. Almost every person is in a position to make money by keeping a few fowls to supply the daily egg needs of the home table, or a greater number to satisfy the appetites of customers or friends. It is a sincere pleasure, not considering their pecuniary value, to be able to have new-laid eggs for breakfast throughout the year. The younger members of a family always relish homelaid eggs, and especially when they are laid by some particular hens that the boys and girls have learned to respect for their good qualities. Possibly the great majority of the eggs that are produced in America are laid by hens owned by the agriculturists, but we are of the opinion that the second greatest number of eggs are laid by the flocks of the village acre, or the small city lot.

To commence with the egg, you will be interested in knowing its composition and formation. Regarding the former, it has been determined that the shells of hens' eggs constitute about 11 per cent, the yolk 32 per cent, and the white 57 per cent of the total weight of the egg. According to tests made at one of the experiment stations, white-shelled eggs have a somewhat heavier shell than brown-shelled. The article, "A Practical Study of Eggs" in this book is one of the simplest and most easily understood explanations of egg formation that we have had the pleasure of reading. It not only tells the manner in which eggs are made, but it explains the reason for deformed and other unnatural eggs and tells how these troubles can be prevented.

A number of terms are used by persons dealing in eggs that are sometimes unintelligible. Frequently an order is sent to the store for fresh eggs, and the eggs that are received are not of the quality that the customer desired. It can be said that the term *new-laid* is used in referring to eggs that have been laid not over three days, while *fresh* is applied by the trade to eggs a week or more old, to storage eggs, or to those preserved in any other manner. Some markets designate new-laid eggs as *fancy fresh* or *strictly fresh*, but when purchasing it is well to be aware of any eggs that are sold as "fresh eggs," because of the broadness of the term "fresh" and the uncertainty of the age or edibility of the eggs.

Eggs are classed among the animal foods, i. e., they contain the same food constituents as are found in the meats and fish and not those that are present in the vegetables; a dinner of eggs and potatoes is just as nutritious as one of meat and potatoes.

One of the striking features of this age is the increased consumption of eggs and poultry. This is largely due to the superior quality of these foods and also to the more attractive condition in which they are now placed on the markets. Then there is the further factor that persons who live in the great centers of population require foods which contain a large amount of sustaining power for brain as well as body, and yet can be more readily assimilated than the heavier foods eaten by those who spend their lives in the open air. Eggs and poultry are especially adapted for this important class.

There is contained in this work, trustworthy information regarding the proper feeding of fowls for heavy egg production. There are two systems of feeding in vogue in America: (a) wet mash; (b) dry mash. The division hinges on the question of whether a mash or mixture of principally ground grain is fed wet or dry. With either system one or more feeds of whole grain are given daily.

The wet mash system has been in use, both in our country and on the continent for hundreds of years, and until recently was thought to be the only satisfactory method of feeding. The mash is fed either in the morning, afternoon or at night, depending on the most convenient time for the feeder, but the quantities of food given at the different periods are not similar. When the mash is fed in the morning or afternoon, it is fed sparingly; if at night, the fowls receive all they can consume. The reason for this change is that a heavy feed of wet mash in the morning tends to make the fowls sluggish and inactive until the meal is digested—an undesirable and unprofitable condition. Successful breeders who practice wet mash feeding, feed a light meal of mash in the morning and two feeds of whole grain in the afternoon and evening. Whether the whole grain or mash is fed at night, this last feed is always a heavy one.

The dry mash system referred to is the feeding of dry ground grain in automatic hoppers that supply a constant amount to the fowls. The advantages of this system over the wet mash is that the fowls are able to eat the dry mash whenever they are hungry. Therefore the system more nearly conforms to the natural gathering of food. If you have watched fowls at liberty you have noticed that they are constantly picking up particles of vegetable or animal matter; they are not able to gather a great supply—to fill their crops—in two or three minutes and then sit down and rest for an hour, but they must keep moving, hunting constantly, or they will go to roost hungry.

Our last year's experience with dry mash feeding has convinced us of its superiority and although the dry mash is in front of the fowls constantly, they eat it by a few mouthfuls at a time and we have noticed little difference in the amount consumed per fowl when compared with the other system. In addition to the mash we throw our fowls one liberal feed of prepared scratching food (mixed whole grains) late in the afternoon and this grain keeps them exercising constantly. There is six inches of fine clover chaff on the floor of the pens and the grain is scattered through this litter. The main reason why we prefer the dry mash system is because we gather 50 per cent more eggs than when we fed the wet foods.

It is not within the province of this introduction to give the formulas of mashes that are used for feeding, but you can make a suitable mash at home by combining palatable ground grains, bran, middlings and blood meal, or you can purchase one of the excellent egg producing mashes that are offered for sale. The latter are mixtures of ground grain, alfalfa (clover) meal and either blood meal, dessicated fish or other animal food. The function of the latter product is to increase the protein content, which in turn is used by the fowl in the formation of the albumen of the egg. Some breeders omit the blood meal from the grain mixture and place a hopper of beef scraps in the pen. You are probably aware that fowls require grit or sharp stones, shell material such as crushed oyster shells and pure water.

Feeding for eggs is carefully considered in this book and we take pleasure in referring you to the articles by Mr. Dryden and Mr. Brooks. These gentlemen are both agricultural workers of known experience and ability, and the information they present is reliable.

The appliances in use in the poultry house are simple and

INTRODUCTION

few in number. In addition to the automatic hopper for feeding the dry mash there are hoppers for the supply of beef scraps, grit and shell, also water fountains for summer or winter. The latter fountain is made with a small kerosene lamp that prevents the water freezing, but its practical value has not been determined.

We constantly receive inquiries from persons who contemplate engaging in the poultry business as to which breed is the greatest egg producer. This question cannot be answered directly for the reason that a breed can be represented by strains of fowls that lay exceptional numbers of eggs during a year, and at the same time, by other strains that are remarkably poor producers. Some years ago a poultryman created quite a sensation in America by developing a strain of Light Brahmas whose females laid more than 200 eggs per year; in fact, some of these females laid as high as 250 and up to 267 eggs in a year. It is a known fact that the usual Light Brahma is an indifferent layer and the estimable work of this breeder substantiates our argument that the strain and not the breed determines whether the fowls will be satisfactory performers.

One of the instructive and interesting chapters of this work is entitled, "Breeding for Eggs." This discusses *pedigree breeding* and advances the information that the greatest egg producer is the fowl that has been bred for the sole purpose of laying eggs. The director of the Maine Agricultural Experiment Station has written an excellent article on the "Egg Yielding Capacity of Hens". Another instructive article is "Better Layers and More of Them" by a poultryman who has developed a strain of heavy laying White Wyandottes.

If you are starting in the poultry business our advice is to select the breed of poultry that you prefer; pay no attention to whether or not the breed is referred to as a heavy layer, but before you place your order, study the advertisements of the breeders and order from one who is making a specialty of heavy egg production. Almost every variety of the more popular breeds has its devotees who are spending time and labor in establishing heavy laying strains and you can secure fowls—egg producers—that will please you.

Your plant for the production of eggs can be of any size. We would emphasize the importance of giving the fowls an abundance of fresh air. Good results are said to be had from covering the front of the house with wire netting and not closing it throughout the year. This demands a special plan of house so that the natural heat thrown off by the fowls' bodies is confined to the rear of the house, and in a measure warms the roosting quarters. It has been learned that fowls can be confined closely with profit—five square feet of floor space to each bird—when eggs for table use are desired; ten square feet per fowl is the proper space when the eggs are to be used for hatching. In *Eggs and Egg Farms* there are references to several money-making egg farms showing the houses in use on the farm; how the chicks are raised and the mature fowls fed and attended.

The colony house system of housing old and young fowls is increasing in popularity and the descriptive article on the Rhode Island egg farms explains this system to good advantage. The Little Compton breeders are making money from their poultry and they all use small colony poultry houses scattered through the fields.

There are a number of markets in which eggs can be sold. New-laid eggs can usually be disposed of among special customers of the home town at a premium over the market price. Attractive card-board boxes with a separate compartment for each egg, called "cartons," can be bought at less than one cent each and one dozen eggs packed and sealed in each box. This is a modern way of selling fancy table eggs and can be used to good advantage and profit. In all cases, new-laid eggs should be graded as to color of shell and size. Do not place white and brown-shelled eggs in one shipment; neither large and small eggs together.

It is preferable to offer new-laid eggs to the grocer in the fall or winter. At this season he can purchase few new-laid eggs and he will handle all you can supply. If the eggs are of high quality, he will retain your supply throughout the spring and summer months, when eggs from the farmers' hens are produced in great numbers. If you are not able to dispose of new-laid eggs at home for a reasonable price, you can ship them to reliable commission merchants in the larger cities. These men will sell the eggs for you on a percentage basis and remit the proceeds. In conducting this business you should have a constant supply of a great number of eggs, so that several cases of thirty dozen eggs each can be shipped every few days. If you have a smaller output we advise your shipping a case a week to a grocer in the city.

Two trustworthy articles, one entitled "The Eastern Egg Market" and the other The "Western Egg Market," present the yearly quotations for strictly fresh or new-laid eggs in the eastern and western cities. These articles explain how the eggs are purchased from the farmers, crated by the dealers and marketed.

The eggs that are held in cold storage and pickled in lime water or water glass are in demand and in fact constitute the main supply during the winter months. These eggs are sold at probably 10 to 15 cents per dozen less than the new-laids.

You can read elsewhere in this work full information in reference to the different systems of egg preservation. The cold storage of eggs is fully treated in an independent article by a writer who has made a thorough study of the business. Of the numerous methods that have been introduced for pickling eggs, the saturated solution of salt and lime is most popular and desirable. The main objection to limed eggs is that a flavor of lime is sometimes detected when the eggs are eaten. Eggs can be preserved in lime water for a year and still be perfectly satisfactory to use for baking. The water glass method of preservation gives better results, but is more expensive. Water glass is a liquid known as sodium silicate and the mixture is made by adding one volume of water glass to nine volumes of distilled or boiled water. The solution is then emptied into an earthen crock and the eggs are inserted as they are laid, or as soon afterward as possible. The writers have referred to the necessity of exercising care in selecting suitable eggs for preservation, and have explained the proper temperature and the most suitable rooms for storing eggs.

A treatise on eggs and the management of egg farms should include a reference to the incubation of eggs, although in this book this branch of the industry has not received prominent attention. For plants of a moderate or large size the artificial method of hatching and rearing chicks is more profitable and labor-saving than the natural.

For detailed information in regard to artificial methods, we take pleasure in referring you to our recent work of this series, *Artificial Incubating and Brooding*—a book filled with valuable directions and practical instructions for buying, setting up, testing and operating incubators and brooders; also our building construction work, *Poultry Houses and Fixtures*; a book which covers the poultry industry in general, *Successful Poultry Keeping*; another which deals with the natural hatching and rearing of chicks, *The Chick Book*; late works illustrated by Artist Franklane L. Sewell descriptive of the Standard requirements of the more popular breeds and varieties of poultry, such as *The Plymouth Rocks*, *The Wyandottes*, *The Leghorns*, *The Asiatics, Ducks and Geese* and *Turkeys*; an inexpensive but authentic work on poultry diseases, *Reliable Poultry Remedies*.

We trust that you will obtain instructive information from *Eggs and Egg Farms*. We wish you the greatest measure of success in the poultry business.

F. C. HARE,

Quincy, Ill., May 1, 1907. *Associate Editor*.

THE EGG INDUSTRY

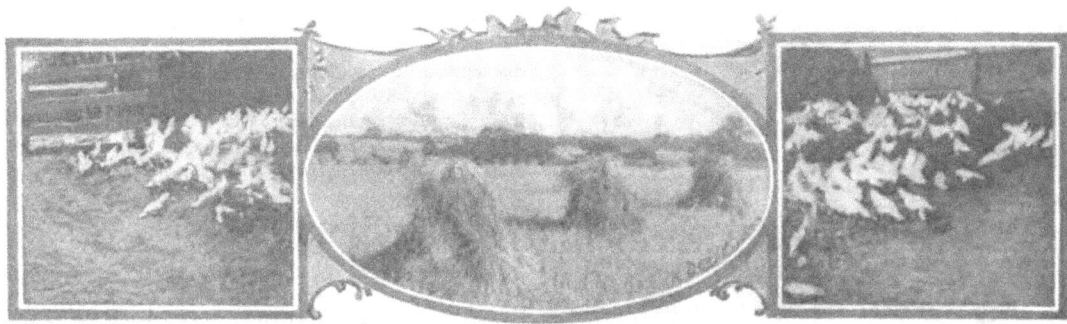

SUCCESSFUL EGG PRODUCTION

GENERAL DISCUSSION OF FACTORS WHICH AFFECT THE POULTRY BUSINESS—
POULTRY KEEPING IS PROFITABLE IF PROPERLY MANAGED—GOOD STOCK
ESSENTIAL — VARIOUS FOODS — METHODS OF FEEDING TO PRODUCE EGGS

ROBERT H. ESSEX

STRIKE the average of figures that have been hashed up again and again to support the contention that there is money in poultry, and you may bank upon it they are in the main correct. There have been many failures on the ragged edge of the business and still they come. The facts are these: It is true there is money in poultry; it is true that the business is an attractive one, and the number of people who are continually attempting to get at the profits, is evidence that they are deeply impressed with some such facts.

A well-known man has said that he made more money from 350 pullets, than from 16 cows. It takes two men and a team to care for the cows and deliver the milk; while one man can easily get around the work connected with the 350 pullets, and yet farmers will keep working away with their cows from four o'clock in the morning until after dark at night and spending on hired help money that might go into their pockets if they devoted equal time and experience to poultry keeping.

Breeding hogs is not nearly so remunerative as breeding fowls. In company with one of the best posted breeders and experts in the country we have visited many farmers' gatherings where the subject has turned upon hog raising. Try as they would expert after expert, they could not show such profit in the business as would warrant continuing it when prices were low. The lecturers themselves "acknowledged the corn," and said they did not pretend that it paid to raise hogs when the market was glutted, but farmers would do it, and they were there to instruct them in the cheaper methods of carrying it on.

It was easy sailing talking poultry after that for there never is such an overproduction of newly laid eggs as to make poultry raising unprofitable.

POULTRY BEATS THEM ALL

Hogs, sheep, cattle, horses, all received attention in the discussions referred to and it seemed to be upon the closest margin of profit, if any, that they were bred.

We remember a farmer who became interested in fowls, and was prevailed upon to give them such care as he gave his hogs. He placed his fowls in the basement of the barn alongside the swine and horses. That was in the late fall. Soon they began shelling out the eggs, and from that time his wife went to town every few days to sell the new laid product. She obtained a good price for the eggs, as she marketed them regularly and the purchasers knew they were to be relied upon. Shortly the farmer began to have doubts about his hogs, and we noticed that upon entering the barn he usually passed the pen of fowls and took up his troubled stand opposite the hog pen. Doubts had arisen in his mind whether or not the hogs were paying, and being an intelligent man, he made up his mind to settle the question. He said he would weigh the hogs and keep account of the food for a month or so. When asked what was his idea, he responded, "Well, I believe these fowls pay me 'way ahead of the hogs, and I tell you what it is, I am going to make sure, and if I find my idea is correct at the end of the month, out go the hogs." It is needless to say the hogs went, and now he is an experienced poultryman, making money at it—that is, he and his wife are.

We could go on multiplying such instances indefinitely, but are satisfied to take it for granted that people now know there is money in poultry, and to rest the case with those who so admirably present their experiences in subsequent pages.

DEMAND FOR EGGS UNCEASING

It is intended in this book to relate what we know—to induce experienced breeders to tell what they know—and to present to you as clearly as possible what you want to know about egg-production. There are two ways to gain experience profitably: First, by beginning on a small scale and growing with the business. Second, by obtaining information under the practical tuition of successful poultrymen.

To either or both of these methods let us add the instruction which can be obtained cheaply by reading the best books on the subject.

The egg market holds firm notwithstanding booms and failures in other businesses. You say, "If it is so remunerative, why is the market not overdone?"

There are two main reasons:

EGGS AND EGG FARMS

1—A GOOD DOZEN

First—The consumer entertains doubt as to the quality of the goods, but will buy an increased amount if convinced that the article is just what is wanted.

Did you ever see a man taking breakfast in a hotel, supremely satisfied that the egg in front of him was perfectly fresh? There is always a doubt of it. Yet good prices are paid for this delicacy (for a new-laid egg is a delicacy even these days), and in most cases the landlord is just as expectant as his guests. Who can be otherwise, when in every grocery and provision store, eggs are quoted at all prices, as new-laid, fresh, strictly fresh, etc. It rests with the consumer to break the shell and see what is what, and sometimes the evidence is not confined to the eyesight.

A reliable man with a reliable egg, has a reliable market, and always will have. The demand increases with the supply.

Second—Egg farming, like every other calling, demands experience. The novice cannot realize this. He makes his investment, then gains his experience, when it is perhaps too late. Although the following little incident has been frequently related, it appeals so forcibly as an illustration that it will bear repetition.

It was on the Atlantic coast, and Mr. A. G. Gilbert, well known as a lecturer on poultry, had impressed upon those present the vast importance of the poultry industry, when a young man in the audience questioned him thus:

"I am anxious to invest $500 in a business undertaking. Would you advise me to engage in the poultry business?"

"Do you know anything about the poultry business?" inquired the lecturer.

"No, sir," was the reply.

"O!" said Mr. Gilbert. "Do you know anything about the drug business?"

"Why, no, sir," was the astonished rejoinder.

"Well, then, my friend," said Mr. Gilbert, "I tell you what it is; I would advise you to tackle the drug business first."

Everybody will at once see that the drug business was chosen as an example of a profession requiring knowledge, experience and care; and it was a happy illustration of the fact that equal industry, experience and care are necessary when about to engage in the poultry business. Does an artist assume to paint before he has mastered the mixing of the colors? Does he beautify his ideal while yet unable to produce the outline? Does the naturalist attempt to enlighten us on the habits of the feathered tribe before he is competent to distinguish the individual types of his pets? No. Yet a young man who cannot tell a cock from a cockerel, a hen from a pullet, a Leghorn from a Brahma, or a sickle from a sythe, assumes to pose as a poultryman, and to know that this most remunerative business is his simply for the asking.

The foregoing are two main reasons why the poultry industry is not overdone.

THE LOCAL MARKET

A report of the Director of Rhode Island Agricultural Experiment Station simply bears out what poultry papers have been impressing upon their readers for so many years. It has this advantage, however—it will be free from any charge of a biased view of the situation. The Experiment Stations are created for the benefit of the people of this country, and every care is taken that none but the most reliable information be sent out. Now let us see what this report states of interest to poultrymen. It applies chiefly to local trade, such as can be worked up in nearly any locality.

The director says: "Another market has come to Rhode Island, furnished by the demands of the wealthy summer visitors and cottagers who desire to obtain and will pay well for fresh fruit and vegetables, dairy and poultry products. Farmers favorably located are learning to supply this special demand. Opportunity is thus offered for largely increased profit if the farmers who cater to this trade will supply the

2—SELECT THE LAYERS AND BREEDERS

The food and care bestowed on this hen were practically thrown away.
She laid only eight eggs during the year.

very best products, prepared according to approved methods, put up in attractive style, and delivered fresh daily at the customer's door. * * *

"In the extreme eastern part of the state the farmers have turned their attention to raising poultry on a large scale. They have in large numbers succeeded so that now their poultry farms are sought by the dealers in poultry and eggs instead of the poultrymen having to seek markets for their products. * * *

"It seems inevitable that there should come in the near future a great increase in poultry keeping in the state. Large areas of land now neglected but well adapted to poultry farming may be economically and profitably turned to this purpose. * * *

"On many a farm the poultry is in reality the most profitable part of the business. This plan should be extended until on every farm adapted to the business poultry is kept to the extent of several hundred fowls. If the farmer himself has no interest in this kind of live stock, he can at least give the son or daughter or wife this opportunity to increase the farm profits, or to gain well deserved pin money."

GATHERING AND SHIPPING TO MARKET

Care must be exercised in gathering the eggs, in packing them, and in shipping to market. The following extract from a prominent authority amply illustrates this fact:

"The egg trade, I am led to understand, has been abused by the farmers, who form the chief source of supply. Their ideas have been to carry to market a full basket of eggs whenever it is convenient to go to town on business. 'Gather them in,' is the policy; fill up the basket, old and young, great and small; good, bad and indifferent; and he is a rascal of a commission merchant who has the audacity to inform the honest farmer that 'ten per cent of your last delivery was bad,' which means ten per cent off the price. The unsophisticated (?) farmer won't believe it, and claims he is being 'done up.'

"A member of the firm of D. Gunn Bros. & Co. had the kindness to illustrate to me, while on a visit to their large establishment, the difficulties they meet in the course of their business. 'In the first place,' said Mr. Gunn, "the farmers will not convey their eggs to us in a proper manner. The great majority of eggs are received in baskets, rattled over a country road for many miles, and naturally many are broken, and more are injured by the jolting and shaking. To illustrate this, come along and see our men candling the eggs. Here is a consignment of eighteen dozen eggs from which three dozen have been taken as being defective. These are called checked eggs, and result from the severe handling they have experienced. The shells are not necessarily cracked, but (holding one before the light) you will observe the yolk has a muddled appearance; it is distributed through a larger portion of the albumen than is the case with this egg, which is perfect. Here is another defective egg, wherein the yolk is so dark that we simply have to discard it altogether.'

"Why, there's a chick of about ten days' growth in that egg," I exclaimed, and sure enough upon breaking it, there were the eyes and blood vessels of the—mongrel, I guess. Several other eggs were broken, some containing chicks, others showing growth of four or five days, but most numerous were the badly shaken yolks.

"Actually," continued Mr. Gunn, "I had to bring a farmer in here to convince him that sometimes we did get bad eggs from him.

The best of these defective eggs, namely, those which have simply been shaken, are sold for, say, two or three cents a

3—EXPERIMENT PLATS FOR STUDY OF BROODER METHODS

dozen less than the prevailing price, and of course the farmer loses a portion of his profit."

COLOR AND SELECTION OF EGGS

"Have you any preference for brown or white eggs," I inquired.

"Well, the brown eggs always sell better; there may be a difference of a cent or two a dozen in their favor, but we seldom receive them all of one color. All sizes and colors are mixed. The farmer has not yet learned that eggs of one color, or assorted sizes, will fetch a bigger price than those of all descriptions, and it is our knowledge of this portion of the business which enables us to make up on good stock what we lose on inferior. Oh, yes, the brown eggs are in greater demand and bring higher prices."

In reply to a query, Mr. Gunn stated that he found brown eggs averaged considerably larger than white eggs. This was the reply received also from Mr. DeLaporte, another prominent commission man. It set me thinking, as my experience is the opposite; and I finally decided that the white eggs they received from the farmers are from the common barnyard stock, which has deteriorated in size, and in egg production, while the brown eggs are from birds which have been improved by the introduction of pure-breds of the American or Asiatic classes. So few farmers keep a high class of the Mediterranean variety that the large white eggs of this class are few and far between. On examination I found only

4—PROFITABLE RETURNS

a very few which might pass for either Minorca or good sized Leghorn eggs.

"The English market," said Mr. Gunn, requires a fifteen pound egg that is, fifteen pounds to the long hundred, or ten dozen. Colored eggs are preferred. A large business might be done there, if I could only obtain the high grade eggs required; but when it becomes necessary to grade eggs from stock having three dozen defective out of eighteen dozen, this trade is simply impossible; otherwise it would be most remunerative to those engaged in the poultry business."

Mr. Park, a prominent merchant, acknowledged that he could give two cents a dozen more for selected eggs (that is, large or brown eggs) than for ordinary stock.

"The market in Toronto last winter," said he, "was very unreliable. From the 1st to the 15th of December, eggs went flying up to forty cents a dozen; then, from the 18th to the end of the month, they were down to twenty cents on account of the great supply.

The following contains some valuable pointers: "The loss in the value of eggs offered in Toronto and other markets through careless handling, is each year considerable. The slightest crack renders the eggs valueless for packing or cold storage purposes, and when sold as "checks" or cracked eggs, from two to three cents per dozen less than standard prices must be accepted. Collected from the nests in a haphazard way and carried to market over rough roads in an ordinary basket, there is usually considerable breakage before the eggs reach the store, where they run the chance of further loss by the handling of the merchant or his assistants. Loss in this way is inevitable so long as proper egg carriers are not used. These egg cases can be purchased at a very nominal figure, say twenty-five cents for a thirty dozen case, and by careful usage will last for years.

"Keep the eggs clean," is the advice which every merchant would impress upon the owners of poultry. An abundance of fresh straw in the hen house is not a heavy expense, and it is essential to a profitable market. If in spite of care the eggs should become dirty, then by no means wash them, as this process removes a glutinous covering from the shell and impairs their keeping qualities."

This advice coming from a responsible firm makes us think. We are, however, of the opinion that the inducement to wash the dirty eggs would be too strong. The uncleanly condition certainly would affect the price of the product.

"Prof. Hilgard, of the University of California, experimented with a view to solving the problem of whether dark or light eggs are richer in their elements. He has made an analysis of the following varieties. Dark Eggs—Partridge Cochin, Dark Brahma, Black Langshan, White Langshan, White Wyandotte, Barred Plymouth Rock. Light Eggs— Brown Leghorn, White Minorca, Black Minorca, Buff Leghorn. No practical difference, in a given weight of eggs, was found in the quantity of waste and edible portions of white and yolk. Chemical analysis yielded a similar result, no differences which might not occur between specimens of the same variety were detected between the several varieties. So far as the examination went, one egg was as good as another, irrespective of the breed which laid it or the color of the shell.

The demand for brown eggs, however, has caused intelligence to be brought to bear on the production of a brown egg layer that might equal the white egg breeds, and a fair amount of success has attended these efforts. The result has been the production of a general purpose fowl, that is, a fowl which meets the requirements of the table and the egg basket, a "go between," as it were. These have been principally bred from crosses of the Mediterranean breeds, layers of numerous white-shelled eggs, upon the larger Asiatic breeds (which lay brown eggs), with sometimes a touch of Indian Game blood. The result of these crosses was an indefinite color in the egg, sometimes white, sometimes brown, sometimes neither. This opened up the way for more experiments, which are being carried on to-day. Some fanciers have taken a general purpose fowl and successfully endeavored to breed a strain of brown egg layers, which can be relied upon; and this tendency has been so much followed that the majority of these breeds now lay a decidedly brown egg. Other fanciers have made endeavors to increase the egg yield of one or more of these breeds, and so successfully, that in individual strains they have been found to be the equal of the Mediterraneans. It

5—PLATS FOR HENS AND CHICKS

has been done by pedigree breeding. Upon that method of increasing the egg yield more will be said later.

The interviews we have referred to suggest that there is, in the egg trade, "room at the top."

The poultryman who will ship, in properly constructed cases, eggs that have been graded according to size and color, can obtain recognition from the commission man, and that quickly. The breeder of pure-bred Leghorns, or Minorcas, which have been selected from an egg-producing strain, can build up a trade at two cents a dozen higher than market price. The breeder of Plymouth Rocks, Wyandottes, Brahmas, in fact any of the brown egg layers can do likewise. The farmer shall not be out in the cold, if he will only use common sense, and grade up his stock by the introduction of pure-bred male birds. Stick to one variety, and purchase a standard-bred male each year, and your common barnyard fowls will lay the golden eggs.

WHY PURE-BRED BIRDS ARE TO BE PREFERRED

Before going into the main elements of profitable egg production, there are a few thoughts we want to share with you, which are really necessary to a satisfactory commencement. They relate to the selection of a breed or variety. It is not merely a question of which is your favorite fowl, but rather for what purpose was that fowl designed, and what are its attributes or peculiarities aside from egg production, because tributary to this branch of the industry is the disposal of surplus cockerels and of hens which have passed the line of profit from an egg producing point of view. Much in this particular depends upon your selection.

Pure-breds should form the foundation of your flock.

Why?

Because they lay better than mongrels; because their eggs may be sold for hatching, and even when disposed of on the market will command a higher price on account of their uniform color; because the surplus cockerels will fetch a good price as breeders, and because being composed of one variety all your pens may receive similar attention and feeding, which economizes labor.

Note what an experienced poultryman says on this point.

"Every farmer who raises common poultry can put money in his pocket this coming season by investing $2 to $4 in a pure-bred rooster. It is as plain as simple arithmetic. Buying a large, vigorous bird, that outweighs your present rooster by two pounds, is equivalent to adding one pound of weight to every healthy chicken you raise next spring and summer. The saying that the male is half the pen applies in this case. The farmer need not pen up his fowls to make this true.

"One pound of weight added to each of the few hundred chickens raised each year on many farms is a big item. The quantity of marketable chicken meat is not only increased by this simple process, but also the quality, for the larger and finer looking fowl, alive or dressed, is easier sold, and at a better price.

"The farmer who wishes to improve his finances will look carefully after just such matters as these. And where the farmer's wife is the 'chicken man' she will do so. It is these strokes that count. Brute force does not hold its own on the farm as well as it once did. The thinking, planning, experimenting farmer is the one who now makes headway and find life on the farm worth living.

"Then there is the important matter of an increased egg-yield. This can readily be brought about with any common farm flock by introducing male blood from the great egg-laying breeds, the Leghorns, Minorcas, Andalusians, etc. A male of this kind, suitable for the purpose, can be bought at a low figure, and he will earn his homestead right several times over by the increased number of eggs his descendants will put into the basket.

"Farmers do not neglect such opportunities as these! With the prices of farm products so low it is wisdom for you to put your thinking cap on and be resolved to improve every chance to better your condition, to earn money.

"Talk the matter over with your wife."

In the show room you may hear visitors exclaim, "Why, my fowls at home on the farm are just as good as these; they are just as big, and I guess they lay as well." Then they'll turn around and say, "Don't you think so, mister?" We tolerate people who talk that way. It is better so. Life is too short to argue with them. If such comments were confined

6—ACTIVE CHICKS JUST FROM THE INCUBATOR

to one or two individuals we might feel inclined to argue the case on the spot; but we have found that there are so very many of these spots, that it is better to overstep them. These people cannot realize how different their birds would look if placed alongside the pure-breds which occupy a coop in the

7—THE FRESH-AIR HOUSE

The Illustration shows the Tolman house as it is used on a poultry plant in Holland.

show room; but they can learn, and if they "mean business," it is to their interest to do so.

Many and many a fancier has taken his medicine right in the show room, and not always in homeopathic doses. Men of experience in the poultry business have received an eye-opener, when placing their birds on exhibition for the first time. "Well, sir," they exclaim, "It beats all, I could have sworn that my birds were bigger, but they seem to dwindle when you get them into a coop alongside these fellows!"

Next, as to the egg producing qualities of pure-breds.

There are hundreds of men in this country who cater to the egg trade, and we know of none who would think of handling anything but pure-breds. They are not in the business for fun, but they cling to the pure-breds every time.

PEDIGREE BREEDING FOR EGG-PRODUCTION

Tributary to the subject of pure-breds arises the question of Pedigree Breeding.

Experiments have been made to see if the number of rows of corn on the cob could not be increased, with success. A similar method to that pursued with the corn is applicable to poultry breeding. For example, one starts with fowls which lay 120 eggs each in a year. Among their descendants are some which lay 150 eggs per year, and these are selected for breeding. From these some are produced which lay 175 eggs per year, and from these, perhaps, the 200-egg-per-year hen is produced. The problem is not quite so simple with fowls as with corn, for it is necessary to breed the males as well as the females, year after year, from prolific layers, in order to succeed. If one looks after the breeding of the females only, he may introduce on the male side blood which is lacking in prolificacy and thus check every attempt at progress. It becomes necessary therefore to breed the males from hens which are varying in the desired direction and which show a cumu-

lative variability in that direction. If the 200-egg bird is to be produced it is just as essential that the male should be from a hen which laid 175 eggs and from a male that was bred from a hen that laid 150 eggs, as it is that the hen was from one that laid 175 eggs and whose mother laid 150 eggs. Improvers of laying fowls are too apt to forget this and introduce males with little regard to their breeding and then wonder why the prolificacy of the flock does not increase.

Attractive! Interesting! Profitable! are terms which can be applied to pedigree breeding. The day is now at hand when no first-class breeder will take any chances on the prepotent quality of his breeding stock, and the fact that there are so many now who follow no system to identify their best breeders accounts in a large measure for the kaleidoscopic rapidity with which poultry fanciers come to the front and just as rapidly disappear from view. It is all chance work with them. They breed a few good birds one year and win. The next year they fail to produce even one, and are at a loss to tell why. The chances are that the sire or dam of the winners went as a martyr to the block, or was inadvertently sold, and perchance caused another man to imagine for a year or so that he, too, had become a successful breeder.

Even plants are affected by selection.

It has been our pleasure to notice divergencies in the color or shape of a flower, and by selecting its seed to inculcate such divergencies in the plants raised therefrom, sometimes with less success.

A leading authority says: "The word selection, taken in its general sense, means choice. In natural history, when applied to plants or animals which man raises under domestication, it assumes a more restricted meaning and is applied only to the choice of individuals considered as agents of reproduction. * * * Much has been said of cultivation as a means of improving plants. The writer believes, however, that the selection of the individual intended to reproduce a sort, has done infinitely more in this direction than cultivation." His conclusion is that "Selection is the surest and

8—WHITE WYANDOTTES LESS THAN SIX MONTHS OLD

most perfect instrument that man possesses for the modification of living organisms."

Prof. A. A. Brigham, Ph. D., Director of the Rhode Island Experiment Station, expresses himself clearly upon this question in its relation to fowls, and what he says will appeal to every intelligent breeder. Coming from such authority it should effectually remove any doubts which exist on the necessity of pedigree breeding, which subject has so far received little consideration by poultrymen. The time is close upon us when the list of winners at poultry shows will be entirely made up from the list of pedigree breeders.

Mr. Brigham says: "The aim in poultry breeding is to produce and perpetuate the best birds, i. e., such as, in their form and other characteristics, answer the purpose of the breeder. The object of the art as usually practiced is to produce the largest quantity of the best quality of certain animal products with the least waste and largest net profit.

"The scope of stock breeding is almost unlimited, and gives ample room for the largest and best, the deepest and highest study and practice of agriculturists.

"When we consider that the excellent improved breeds of poultry of the present day all originated from wild and slightly esteemed progenitors in the past, we gain some idea of the amplitude of opportunities which are open to the intelligent breeder in the present condition of poultry culture.

"Success in stock-breeding depends upon the certainty that the progeny will at the time of procreation, inherit the general or mingled qualities of the parents. Color and outward shape, and the characteristics and tendencies of the internal organs and structure are thus transmitted. Heredity extends even to the transmission of longevity, fecundity, disposition and habits.

* * *

"Prepotency is increased by carefully breeding together animals of like inclination, and the longer this process is continued the more certainly is the peculiarity transmitted. Prepotency in a given respect may thus come to be a quality of a flock, of a family and of a breed.

"The principle of prepotency is of special value to the poultry breeder, in that it enables him to select for his particular purpose, an adapted breed which is, as a breed, prepotent in the line desired; next, to obtain as foundation stock individuals from a family known to be especially prepotent in the particular respects desired; and, finally, to use as the sire of his animals' offspring a male individually prepotent and most certain to transmit the desirable characteristics to his get. Prepotency is of great practical utility in crossing males of pure breeds upon common stock, to rapidly improve the poultry of a country or district. * * *

"It is a principle in stock-breeding that coupling two animals possessing the same quality, defect, or disease, will tend to fix and intensify that quality, defect, or disease in the offspring.

ESSENTIAL CONDITIONS FOR PEDIGREE BREEDING

"Two essential conditions must invariably attend successful inbreeding, viz , sound constitution and perfect health. With these as a foundation, close inbreeding may be practiced with the best of results, as shown by the breeds thus produced and perpetuated by successful stock-breeders.

"The qualities of fattening easily and quickly, of early maturity, of enormous egg production, all have been brought to the highest perfection in individuals and families which have resulted from close inbreeding. * * *

"From the study we have made of the principles of breeding we must conclude that the ancestry of our breeding animals is of very great significance in determining results. The development of the best breeds of the farm stock has been in a very striking degree the fixing of the desirable and prepotent characters of a few ancestors upon numerous descendants.

"The pedigree of an animal is his line of descent, his list of progenitors, in short, his ancestry. * * *

'The perfectly prepared pedigree of an animal shows the foundation stock of the breed that enters into his line of ancestry, and then step by step exhibits the different links in the

9—AN ENGLISH PORTABLE HOUSE ON THE MOVE

chain of life, indicating to what extent the foundation stock and their progeny re-enter the pedigree, the closeness of interbreeding, the outbreeding, the use of unknown, doubtful, or undesirable sires or dams, in fact, all the blood relationships of which the animal is the result " * * *

"Thus, gradually, a type is developed, which, after several generations, becomes fixed and certain of transmission to continued generations. What the family type shall be depends upon the environments and the standard fixed in the mind of the breeder, who selects, at first, and continuously, those animals which come nearest to his ideal, or possess qualities which he wishes fixed in his stock. He breeds them together in accordance with the laws of heredity and variation, continuing a process of selecting and discarding according to his fixed ideal or standard of excellency, and gets his desire."

That's just it—"gets his desire," and the poultry fancier's desire is to be at the top. By adopting this method of breeding he is facing in the right direction to get there. The egg farmers desire is that his hens shall lay 200 eggs a year. He adopts this method of breeding and they lay 200 eggs a year;

but as this trait has yet to be generally established, it means years of endeavor, although every year's results will spur him on to greater efforts.

QUALITY—NOT QUANTITY

In this age of so-called over production it should be plain to everybody that what is wanted is not quantity, but quality. We have too much, no doubt, of inferior goods, but quality

10—WHITE LEGHORN HENS

always brings good prices, and sells at a premium. While there is no over-production in poultry and eggs, there is a decided difference in price. Quality governs.

Mr. C. H. Wyckoff tells of a case which illustrates the point.

"Two years ago the plum trees in his poultry yards were fairly matted with young fruit. There was more fruit by far than the trees could ripen into large, showy plums of good flavor, or, in other words, into salable fruit." Mr. Wyckoff knew that the wise thing to do was to thin out the crop. This he did, assisted by his wife and hired man. They did the work at odd times, keeping account of the number of hours each worked at this task. The three put in what was equivalent to thirteen days of one person's time. They simply used scissors, thinning out from one-third to one-half of the crop. Said Mr. Wyckoff: "The result was this: while my neighbors and others hereabouts were peddling their plums in Groton at 30 and 40 cents per bushel, I sold mine without any effort at $1.50 per bushel, and they were well worth the difference in price. They were uniformly large and of good flavor, while those that had been permitted to grow wild were half size and hardly fit to eat!"

Along this line another writer says. "Experience shows that quality must be the watchword if we are to do a profitable business in fruit, in poultry, in almost anything. It costs extra effort, it calls for hard work, but it is still the short and pleasant road to success. Referring to the quality of dressed poultry, we some time ago had a talk with a Mr. Allen, a celebrated dressed-poultry dealer. He said to us that he was then getting twenty-eight cents a pound for chickens. We looked up the market quotations and found chickens, dressed, quoted at twelve to thirteen cents per pound. The difference in price was due to quality. Quality is what pays. Mr. Allen said that he bought from certain poultry raisers whom he had years ago carefully instructed what to raise and how to raise it, who always brought him chickens of a grade that other dealers could not duplicate, and these he never had any difficulty whatever in selling at the highest price. It was the poor, cheap stuff that did not pay. For the best there was always a demand beyond the supply. The general law of supply and demand operates uniformly the world over, where it is not interfered with by unwise legislation. If the demand exceeds the supply, so that the customers bid against each other, the price will go up; but if the supply exceeds the demand, so that the sellers bid against each other, the price will go down. It makes a great difference which class does the bidding, but always quality pays, for the supply of the best is always inadequate to meet the demand for the best."

HOUSING THE FOWLS

Assuming that you have sufficient land for your purpose, the number of fowls to be kept should be governed by the house accommodation you can afford. That is where many poultrymen fail. This year's stock is bundled into last year's houses, although there may be half as many again fowls. This won't do; it is the commencement of trouble. Lice, disease, puny stock, and continual worry, with an accompanying proportion of non-success, may be attributed, in many cases, to want of house room. Therefore limit your stock to the expenditure in this direction. There is no need to build elaborate poultry houses; the simpler they are the better. Study your climatic requirements, then put up houses which may be well aired, and which are so conveniently arranged as to economize labor, and provide accommodation for your every requirement in all seasons.

11—WHITE LEGHORN PULLETS FOUR MONTHS OLD

You should have pens for the chicks when they leave the brooders, and furnish them sufficient house room to prevent crowding. Overcrowded chicks never attain health or size;

THE EGG INDUSTRY

and even if they should happen to come along well, it means so much extra care and attention. You must allow for the separation of the cockerels from the pullets before they attain maturity. You should provide pens for fattening the surplus cockerels, as they will need to be fed separately. You must allow for breeding and laying pens, and lots of room for exercise during the time the fowls are confined to the house. Pens or coops for surplus males should be made, and it is well that they should be able to exercise themselves. Then, if you think of placing any of your birds in a show room in the best possible condition, there should be coops in which you can prepare them.

The room for your food should be rat proof, and of course the hen house, too, if you can afford it. It is advisable to have solid wooden partitions in your house at certain distances, if your house is a continuous one—at least every half dozen pens.

FRESH AIR PREVENTS DISEASE

As to ventilation, we do not place much faith in the many appliances that some poultrymen feel bound to have. Given a house where we can open our windows and doors during the fine weather, and we want no patent ventilating aparatus. In nearly all cases where ventilators are used, there arises a question whether or not the houses are properly aired, and in many cases a draft is created. This is detrimental to the health of the fowls.

One of the chief preventives of disease is fresh air, and it can be better introduced by opening all the windows of the house than by means of any ventilator that has ever been invented. If the houses are kept well cleaned the necessity for ventilation is reduced fifty per cent, that is to say, the foul air is not allowed to accumulate so rapidly.

Much depends upon the location of the houses. "High and dry" is a good motto in this connection. Some poultrymen prefer a long building of continuous pens, others what is called the scratching shed plan, which provides for an open shed for every pen. In all your constructions bear in mind the necessity for time saving in attendance upon the fowls—economy of labor; have as few nooks, corners and uneven surfaces as possible; provide for natural warmth at night in winter; have no fixtures that you can dispense with; make the nests, roosts and droppings boards movable; windows not too large; floors as you choose, boarded or not, according to circumstances, but they must be dry; in fact, keep in view economy, cleanliness and the health of your stock, and never mind the frills.

Damp, frosty walls in hen houses have troubled many a poultryman. The open, well-aired house is one of the best preventives of this annoyance. The house that becomes warm during the day and cold at night will have damp walls just as sure as a dirty house will harbor lice. If you are bothered with damp walls and do not care to open your windows or other ventilators continually, the only way to dry them will be to light a fire periodically as the houses require it; this method however is dangerous and generally results in colds.

FEEDING FOR EGGS

The young person in the poultry business is apt to become discouraged by the complicated suggestions he reads on the food question. The more he reads the more discouraged he becomes. Don't let a little thing like that throw you off the track. All the rations you read about are good, that is, all those that are recommended by reliable publications. The kind of food a hen should get, the amount she should consume, and the time she gets it, should be governed entirely by circumstances. What will cure a horse, will kill a man, and for that matter, what will kill one man may not materially

12—AN EXPERIMENTAL DOUBLE HOUSE IN ENGLAND

affect another. Sometimes it depends upon the strength of the man, and sometimes on the strength of that he imbibes. And so it is with fowls. A Leghorn and a Brahma should be fed differently. A fowl on a big range must not be fed similarly to one that is confined to a house. Food that produces eggs in winter will be considered heavy feeding in summer. These are things the novice has to learn.

BALANCED RATIONS

Do you not think that breeders are beginning to dabble into scientific feeding? Certainly they are; it is one of the several methods intelligent men are using to increase the egg production of their flocks. The cattle men have been at it for years endeavoring to increase the supply of milk, and with very beneficial results. They feed a balanced ration which possesses, as nearly as possible, all the forms of nourishment that enter into the composition of milk and possesses them in like proportion. Some poultrymen are working upon similar lines, taking the composition of an egg as their basis.

The experiment Stations are doing good work, and breeders will await the result of their investigations with interest. Some of the results will be useful in determining the food question, but they are not of necessity so.

We have in mind an experiment that was conducted over a period of seven months to discover the effects of nitrogenous as compared with carbonaceous food upon fowls. One group was fed heavily with corn, the other with wheat—both with a proportion of other grain. We will, for simplicity, call them corn fed and wheat fed fowls respectively.

13—A LAYING HOUSE OF MODERN CONSTRUCTION

The conclusions arrived at were as follows:

(1) The wheat fed fowls gained 354 pounds, while the corn fed fowls gained only 34 pounds during the same time.

(2) The wheat fed laid 17,459 eggs, the corn fed only 9,709.

(3) A larger per cent of the eggs laid by the wheat fed fowls were fertile, the corn fed laying many infertile eggs.

Such experiments as these would have a far-reaching effect on the poultry industry, if the conclusions were generally accepted. They would seem to prove that nitrogenous feeding is away ahead of carbonaceous feeding, but they prove nothing of the kind, when we come to analyze them.

Take the first case for example, which says that the wheat fed fowls gained 354 pounds in seven months and the corn fed gained only 34 pounds. Any practical poultryman when he considers the ration will be surprised to find that the corn fed fowls had any gain at all credited to them at the end of the seven months. Fancy a fowl thriving on potatoes, corn and oats. There was, we admit, a feed of clover in the third, fourth and fifth months and some wheat screenings during the first, second, sixth and seventh months, the last two being most important of all, there was nothing but corn, potatoes and oats, and during the fifth month they existed on corn and potatoes with some clover hay thrown in, while on the sixth month the poor things were fed corn and potatoes alone.

On the other hand see what the wheat fed fowls reveled in every month—potatoes, hominy, feed, brown middlings, corn, oats and fresh bone; wheat screenings for three months, clover hay for three months and oil-cake for two months. They also received every month from 200 to 450 pounds of food more than the corn fed fowls. The chances are that the corn fed fowls got sick of their rations and would not eat. The fact is, instead of making any gain in flesh they lost regularly during the last four months of the experiment, which proves they were not getting properly balanced rations.

As to the second conclusion: If a poultryman were to feed his fowls as these were fed, he would not expect any eggs, and the fact that those experimented upon were losing flesh during the last four months, should have shown the persons in control that something was wrong in their manner of feeding. If the fowls were not vigorous it is natural to suppose their eggs would be infertile. The conclusion arrived at states that, "Although the nitrogenous ration costs slightly more, yet it was more profitable, because more eggs were laid and the fowls gained more in weight. The eggs from the nitrogenous fed fowls were larger, more fertile and hatched better and produced far more vigorous chicks than those laid by hens fed on carbonaceous rations. Both lots of fowls remained in a healthy, vigorous condition during the entire test."

It has been impressed upon poultrymen again and again that fowls need a variety of food. It has been proven that fresh cut bone is one of the best egg producers in existence, and yet somebody argues that because corn fed—ill-fed—fowls do not lay as many eggs as fowls that are fed wheat and cut bone, therefore carbonaceous feeding is comparatively undesirable.

Our conclusion is, that incorrectly balanced rations, or those that contain no green food, and no animal food, are dear at any price. Further, the fact that eggs become smaller is sometimes one of the signs of unthrifty fowls. Again, fowls which lose weight during the last four months out of seven, cannot be in a healthy state, unless at the beginning of that period they were far too fat. Furthermore, it is not wise to adopt on the spur of the moment the conclusions based upon experiments which were not conducted upon parallel lines.

GREEN FOOD IMPORTANT

Among the valuable experiments that have been conducted are several on the value of clover as a food. At the Kansas Experiment Station, Alfalfa, (Lucerne) was tested as food for hogs, and proved pretty clearly the advisability of using green food, if only to promote digestion.

The hogs were divided into four lots of ten each.

We quite: "Lot 1 was fed dry Kaffir-corn meal and alfalfa hay, lot 2 whole Kaffir-corn, lot 3 dry Kaffir-corn meal, and lot 4 wet Kaffir-corn meal. The alfalfa hay was of the

14—INCUBATOR CELLAR WITH LABORATORY ABOVE

best quality and carefully cured. It was dry fed in a large feeding trough. The pigs were confined in large pens with open sheds. The test began November 24, 1898, and covered 9 weeks. Lot 1 gained 90.9 pounds or 10.88 pounds per bushel of dry corn meal and 70.83 pounds of alfalfa; lot 2 gained 59.4 pounds or 8.56 pounds per bushel of grain; lot 3 gained 52.4 pounds ar 7.48 pounds per bushel of grain; and lot 4 gained 63.3 pounds or 8.09 pounds per bushel of grain. These results are not due to the feeding value of the alfalfa alone, but also to its influence in aiding the hogs to better digest the Kaffircorn. The alfalfa hay also gave a variety to the ration, making it more appetizing and inducing the hogs to eat more grain. * * * The hay fed hogs ate more grain and gained more for each bushel eaten.

"In a former experiment at this college pigs were pastured through the summer on alfalfa with a light feeding of corn. After deducting the probable gain from the corn, the gain per acre from the alfalfa pasture was 776 pounds of pork."

We do not wish to suggest that the digestive organs of hogs and poultry are alike; they differ materially, but we quote this instance in order to indicate the fact that a costly concentrated food may be made more nutritious by the addition of a bulky food.

Those breeders of poultry who have used alfalfa would not be without it, and there is every reason to suppose that its use results as satisfactorily as in hog feeding. We are not to

We are not sure that the final conclusion of the experimenter is quite correct. Referring to the last experiment, in which the use of animal meal resulted most favorably, he says: "The results have been twice favorable to bone, and twice to animal meal, but this last experiment is more decisive than any preceeding."

If that be so, we do not understand the facts We have before us details of two of the experiments, including that from which we have quoted. In the test wherein the meal fed fowls came out ahead, the cost per egg from them was about three-quarters the cost of those from the bone fed fowls; while in the test wherein the bone fed fowls were winners the eggs from them cost only about one-half the cost of those from the meal fed fowls. So that as between the two most decisive experiments, cut green bone finished considerably in the lead.

FLAVOR OF EGGS

Another point which should not be overlooked is the effect of certain foods upon the flavor of eggs.

15—A GASOLINE HEATED COLONY BROODER AND CHICKS

imagine that alfalfa is the best green food in existence, but reckoning on its cost compared with other foods, we are not far astray in saying it is the cheapest. Cabbage, for instance, contains more nutriment than alfalfa, and, fed in equal quantities, will likely be a better egg producer, but it is a much more expensive food.

To obtain best results alfalfa should be cut between the periods of medium and full-bloom, and should be carefully cured, bearing in mind that the leaves are most nutritious. We cannot place too much importance on the feeding of green stuff. It goes a long ways towards successful breeding.

CUT BONE VERSUS ANIMAL MEAL

Sometimes we jump at conclusions too quickly altogether. Experiments, even when conducted on a proper basis, are not always conclusive. Cases in point are investigations which have been made regarding the comparative value of cut bone and animal meal. Of four experiments, two resulted in favor of the bone, and two in favor of the meal. Now, if a reader had known the result of only one of these experiments, he would have sworn by it, of course. Therefore, beware.

It seems to be a fact that if hens are largely fed upon highly pungent foods the eggs will be to some extent tainted by those foods. A fishy diet will impart a fishy flavor; onions will give some of their pungency to the eggs. We have read of fowls eating the carcass of a dead muskrat and laying eggs with a musky flavor. All of the instances which have come under our notice where eggs have been affected unfavorably by the food have been cases where the food possessed a strong odor and had been consumed in considerable quantities. Other foods that have been equally disgusting to our sense, but which lacked volatile properties, have seemed incapable of noticeably affecting the flavor of the eggs.

INCUBATORS AND BROODERS

It is absolutely impossible to compete with the present-day breeders who raise poultry and eggs for market unless you use incubators and brooders. To begin with, the stock cannot be bred in sufficient numbers, and again, just at the time when you want to place your eggs in incubation you are at a standstill if you depend upon hens; the result is the produce is not hatched at the proper season, and therefore cannot be placed on the market when high prices prevail, and it is by marketing at such times that the poultry business returns enormous profits. Unless you compete at such periods you might as well drop clear out of sight.

The breeder of a few fowls needs no incubator if he has a variety of fowl which incubates. On the other hand, if he breeds Minorcas, Leghorns or any of the non-sitting breeds, he either must have an incubator or purchase hens for sitting. Our experience is that when the number of chicks raised exceeds a hundred or so, the incubator saves time, trouble and money.

LEADING BREEDS OF FOWLS

It is better that you should confine yourself to one breed; you will be more successful, and like it better; you can pay more attention to perfecting that breed, and you will feel as happy as a man with "just one girl."

In the poultry kingdom there are egg producers, market fowls, general purpose fowls, and fancy fowls.

The egg producing field, until late years, has been monopolized by the Mediterranean breeds; so called by reason of their origin on the north shore of the Mediterranean Sea. These birds have been bred purely for egg production and they all lay white eggs of various sizes. Other fowls that have been carefully bred with a view to increased egg production have given satisfactory returns, and as we have said elsewhere there are individual specimens that equal the Mediterranean breeds. We will refer briefly to the principal breeds and leave the reader to take his choice.

THE BARRED PLYMOUTH ROCK

The Barred Plymouth Rock is nearly too old to need description. Nearly every farm in the country has had this variety at some time, and the reader will readily recognize them by the following brief description. During the past ten years there have been more Barred Rocks than any other breed placed by farmers upon the roaster and broiler market. They are good winter layers.

16—PICTURESQUE DUCK YARDS ON AN ENGLISH ESTABLISHMENT

Plumage—A bluish gray, barred with a very dark blue, which approaches black.

Comb of All Varieties of the Plymouth Rock—Single, and comparatively small.

Standard Weight of All Plymouth Rocks—Cock, 9½ pounds. Hen, 7½ pounds. Cockerel, 8 pounds. Pullet, 6½ pounds.

THE WHITE PLYMOUTH ROCK

The White Plymouth Rock is a more recent variety, and is claimed to be a sport (or accidental production) from the Barred Rock. It is similar to the Barred Rock in every respect except color. They are not so extensively bred, but have an advantage, as a table fowl, over the Barred, in the absence of dark pin-feathers which disfigure the first named variety when carelessly plucked. The greatest difficulty experienced by fanciers is to eradicate the creamy shade from the plumage and at the same time preserve the yellow legs and beak. Every year his trouble is becoming less noticeable, so much so that at the best shows few creamy birds are now seen. As layers they equal the Barred variety.

Plumage—White.

THE BUFF PLYMOUTH ROCK

The Buff Plymouth Rock is the most recent addition to this general purpose breed and has in a short time become very popular. It is claimed that the plumage assists in producing and maintaining the much desired yellow skin of a table fowl. It will take some years before the black and white in wings and tail, which breeders have to contend with, will totally disappear. Pure buff birds are extremely valuable. All the Rocks excel as winter layers, and this variety is no exception to the rule. Plumage—Golden buff.

THE SILVER LACED WYANDOTTE

The Silver Laced Wyandotte perhaps matures quicker than the Plymouth Rock, but at maturity averages about a pound less. The difficulty to be encountered is the breeding of well defined lacing which, when obtained, renders this fowl particularly attractive. It is of a more blocky build than the Plymouth Rock; makes a good table fowl, and is its equal as a winter layer. It has a rose comb, which should fit closely to the head, and should not be too meaty. The dark pinfeathers appear in this variety, as in the Barred Rocks.

Comb of All Varieties of the Wyandotte—Rose and low.

Plumage—Black and white, distributed as illustrated in the Standard of Perfection.

Standard Weight of All Varieties of the Wyandotte—Cock, 8½ pounds. Hen, 6½ pounds. Cockerel, 7½ pounds. Pullet, 5½ pounds.

The Golden Laced Wyandotte is similar to the Silver Laced Wyandotte in every respect, except that golden bay is substituted for white in the plumage. It makes an equally handsome bird.

THE EGG INDUSTRY

THE WHITE WYANDOTTE

The White Wyandotte has had a big run among the broiler men. Rapid maturity, absence of dark pin-feathers, and a blocky little frame quickly filled out, has made it possible to place a meaty morsel on the market at twelve weeks. As a winter layer it equals the other varieties of the Wyandottes, and has been bred extensively by the fancy, who have been working hard to retain the yellow legs and skin while breeding out the creamy tinge in the plumage.

Plumage—White.

THE BUFF WYANDOTTE

The Buff Wyandotte is one of the most recent productions in the Wyandotte class, and the demand for it has increased very rapidly. It has the same advantage of color as that claimed for the Buff Plymouth Rock. Like the other Wyandottes, it matures quickly, and is being extensively tried for broilers. It makes a good winter layer. The black and white in wings and tail will take some years to breed out, but perhaps this lends additional zest to the pursuit by the fancier.

Plumage—Golden buff.

PARTRIDGE WYANDOTTE

This variety, with its cousin, the Silver Penciled Wyandotte, is the most recent addition to the Wyanandotte class. Its plumage is similar to that of the Partridge Cochin, a much older breed. Being a new breed, there is of caurse some difficulty in breeding it up to standard requirements.

LIVER PENCILED WYANDOTTES

One of the most beautiful of the Wyandottes; admitted to the standard in 1903. The plumage is similar to that of the Dark Brahma. Difficulties in breeding to standard requirements will arise for some years.

The Black Wyandotte is similar to the other varieties, except that it is black throughout, and perhaps not as desirable as a table fowl for that reason. In egg production there is little difference.

THE BROWN LEGHORN

The Brown Leghorn is a favorite family egg producer. A sprightly, ever scratching bird, which lays perhaps as well as any variety, though the eggs are on the small side, except in individual strains.

Comb—Single, rather large.

Plumage. Male: Breast, black; back, red, striped with black; neck and saddle, brilliant red, with black stripes. Female: Breast, salmon; back, brown, penciled with darker brown; neck, yellow, with black stripes.

The Rose Comb variety is identical, with the exception of the comb, but there has existed a difficulty in getting them up to size.

There is no standard weight for Leghorns.

THE WHITE LEGHORN

The White Leghorn is identical in size and shape with the Brown; it lays just as well, but the eggs are larger as a rule. More of these birds are being bred for laying purposes than in former years. Many of them are used as a cross on the larger varieties to produce broilers, but this method has not been commonly adopted. The creamy shade of the plumage invades this variety, as it does the other white birds.

Comb—Single, rather large.

Plumage—White.

The Rose Comb variety is identical, with the exception of the comb, but the majority of the birds are slightly smaller, and lay smaller eggs than the single comb birds.

17—A "RANGE" OF PORTABLE HOUSES MUCH USED IN ENGLAND

THE BUFF LEGHORN

The Buff Leghorn is a comparatively recent addition to this breed, and has not yet generally acquired the sprightliness, shape, and style of the other varieties. It should run the Whites close in the competition for popularity, by reason of the favorite color of its plumage They are larger in many cases than the Browns or Whites, and compete with the Whites in the size of their eggs. In the race for perfect plumage, it has jumped to the front of the Buff breeds.

Comb—Single, rather large.

Plumage—Golden Buff.

BLACK AND SILVER DUCKWING LEGHORNS

Black Leghorns and Silver Duckwing Leghorns are less extensively bred than any of the other varieties named. The Blacks are in advance of the Duckwings in this respect. The plumage of the Black is as its name indicates; while the Silver Duckwing male's hackle is silvery white with black stripes; back, white; breast, black; tail, black. Female—Hackle, silvery gray, with narrow black stripes; back, light gray; breast, light salmon; tail, black or brown, becoming gray in the top feathers.

Comb—Single in each case.

EGGS AND EGG FARMS

THE ANDALUSIAN

Or, as it is frequently called, Blue Andalusian, is another of the egg producers not very widely bred. In size it lies between the Leghorn and the Minorca. Andalusians are first-class layers. They are becoming more popular. It is a sur-

18—WHITE LEGHORN PULLETS LAYING AT SIX MONTHS

prise that more have not been bred, but probably this is because so great a number of the chicks are culls in color, running very light.

Comb—Single, rather large.

Plumage—A slaty blue, laced with a darker shade. In the male, the neck, back, saddle and tail approach black.

Standard Weight—Cock, 6 pounds. Hen, 5 pounds, Cockerel, 5 pounds. Pullet, 4 pounds.

THE BLACK MINORCA

The Black Minorca is a favorite breed, combining size with production of large white eggs. During the last ten years a great increase has been made in the size of the comb, which, on exhibition specimens, is now required to be quite large. This renders it difficult to preserve it from frost bite in severe climates, and therefore the egg production is affected in extremely cold weather. In warmer seasons they are unequalled as egg producers.

Comb—Single, very large.

Plumage—Glossy black, with a green tinge.

Standard Weight, S. C. Black—Cock, 9 pound. Hen, 7½ pounds. Cockerel, 7½ pound. Pullet, 6½ pounds.

R. C. Black and S. C. White—Cock, 8 pounds. Hen, 6½ pounds. Cockerel, 6½ pounds. Pullet, 5½ pounds.

The White Minorca is similar in every respect to the Black, with the exception of color, though a difficulty has been experienced in keeping them up to size. Prominent breeders are now overcoming this drawback.

Comb and Standard Weight—Identical with the Black variety. Plumage—White.

The Black Spanish has not as many friends as formerly. The impression is gaining ground that continual breeding for the white face has undermined the vitality of the breed. However, it may be that the difficulty of obtaining this pure face accounts for the lack of breeders of the bird. In egg production it resembles the Minorca.

Comb—Single, rather large.

Face—White.

Plumage—Glossy black, with green tinge.

Standard Weight—Cock, 8 pounds. Hen, 6½ pounds. Cockerel, 6½ pounds. Pullet, 5½ pounds.

THERE ARE OTHERS

There are yet other fowls from which a choice might be made, but as a class they have little reputation as egg producers, although individual instances of prolificacy in this particular will appear in this book. The question of breeds will have been gone into deeper than was intended, but, having "taken hold of the plow," it is deemed wise to omit no fowl which might by any chance be accepted as an egg producer such as might be satisfactory to the egg farmer. Birds that are omitted are either distinctly non-utility breeds, or are inferior to those which have been named in their respective classes.

THE ASIATIC CLASS

The Asiatics are considered to be table fowls rather than egg producers. They are of great size and mature slowly. They possess feathered legs to a greater or less degree, and are a fluffy, full-feathered class, composed of:

Brahmas (Light and Dark); Cochins (Buff; Partridge, White, and Black); and Langshans (Black and White).

Standard Weights—

Light Brahmas—Cock, 12 pounds. Hen, 9½ pounds. Cockerel, 10 pounds. Pullet, 8 pounds.

Dark Brahmas—Cock, 11 pounds. Hen, 8½ pounds,

19—A REED POULTRY HOUSE COMMON IN HOLLAND

Cockerel, 9 pounds. Pullet, 7 pounds. Cochins—Cock, 11 pounds. Hen, 8½ pounds. Cockerel, 9 pounds. Pullet, 7 pounds.

Black and White Langshans—Cock, 10 pounds. Hen, 7 pounds. Cockerel, 8 pounds. Pullet, 6 pounds.

THE EGG INDUSTRY

THE FRENCH CLASS

These are composed of table fowls, pure and simple. Of course they lay eggs, but, as in the Asiatics, it is only individual strains that distinguish themselves in this particular. The class consists of Houdans, Creve Coeurs and La Fleche.

Standard Weights—

Houdans—Cock, 7 pounds. Hen, 6 pounds. Cockerel, 6 pounds. Pullet, 5 pounds.

Creve Coeurs—Cock, 8 pounds. Hen, 7 pounds. Cockerel, 7 pounds. Pullet, 6 pounds.

La Fleche—Cock, 8½ pounds. Hen, 7½ pounds. Cockerel, 7½ pounds. Pullet, 6½ pounds.

THE ENGLISH CLASS

This class is made up of ideal table fowls, that is, from an English point of view. The skin is white. In egg production they are about equal to the French class. The following are the varieties contained in this class:

Dorkings—White, Silver Gray, and Colored.

Standard Weights—

White Dorking—Cock, 7½ pounds. Hen, 6 pounds. Cockerel, 6½ pounds. Pullet, 5 pounds.

Silver Gray—Cock, 8 pounds. Hen, 6½ pounds. Cockerel, 7 pounds. Pullet, 5½ pounds.

Colored—Cock, 9 pounds. Hen, 7 pounds. Cockerel, 8 pounds. Pullet, 6 pounds.

Buff Orpington—Admitted to the American Standard in 1903. A popular general purpose fowl, but somewhat larger than the American breeds of that class. Standard weights: Cock, 10 pounds. Hen, 8 pounds. Cockerel, 8½ pounds. Pullet, 7 pounds.

THE GAME CLASS

With the exception of the Indians and Malays, this class is purely an exhibition collection, and possesses no standard weights.

The Indian is a first-class table fowl, and has been freely used as a cross upon other breeds to improve their size or quality. It is a fair layer. There are two varieties, Cornish and White Indians.

Standard Weights—

Indian—Cock, 9 pounds. Hen, 6½ pounds. Cockerel, 7½ pounds. Pullet, 5½ pounds.

Malay—Cock, 9 pounds. Hen, 7 pounds, Cockerel, 7 pounds. Pullet, 5 pounds.

THE HAMBURG CLASS

The Hamburgs are a small, rose comb breed, of great egg producing capacity, but their eggs are, in most cases, too small to be of use for market purposes. The class consists of:

20—THE TOLMAN HOUSE AS USED IN WASHINGTON

Black, Golden (Penciled and Spangled); Silver (Penciled and Spangled), and Whites.

No standard weights are allotted to them.

EXERT EARNEST EFFORT

And now, having offered such suggestions as have occurred to us as being necessary to your welfare, we advise you to study carefully, and not hastily, the experiences of the good men, who, for your benefit, have given of their knowledge in the subsequent pages of this book. Overlook not even the smallest items—these are often of the greatest importance—then, having the benefit of their many years' accumulation of knowledge, it simply remains with you to profit by it.

Do not consider the poultry business a ready made recreation, but enter it with such a vim, with such earnestness as will assuredly present to you that which is most highly prized by every man who is a man—independence.

DEMAND AND SUPPLY

THE FARMER'S OPPORTUNITY

THE MARKET FOR EGGS IS INEXHAUSTIBLE—IMMENSE FOREIGN AS WELL AS DOMESTIC DEMAND—SMALL CAPITAL REQUIRED TO BEGIN—THE POSSIBILITIES OF THE BUSINESS UNLIMITED—METHOD AND PATIENCE WILL BRING SUCCESS

JOSEPH A. TILLINGHAST

I HAVE been asked to give you some practical thoughts on "Poultry on the General Farm." The importance of this subject must be acknowledged when we consider that with all the specialists in poultry culture, we must still look to the general farm for a large part of our supply. That you may not fear over-doing the business, at least for a little time, and to show the extent of the industry, let me give you a few figures. The dairy products of the United States for one year amounted to $254,000,000. We are in the habit of looking at this branch of farming as one of large extent, but we find the poultry products for the same year to be $560,000,000, or more than twice as much and still not enough, for during the same year 13,000,000 dozens of eggs were imported, and the total value of poultry and eggs imported was probably $20,000,000.

IMMENSE FOREIGN TRADE

This $20,000,000 ought to have been jingling in the pockets of American farmers and poultrymen rather than to have been sent to foreign countries. Even our little state of Rhode Island used from outside of the state about $800,000 worth of eggs. Great Britain imported eggs and poultry to the value of £5,675,000, or $27,637,250. London alone used other than English eggs to the value of $6,915,000. France reckoned the value of her poultry products at $77,920,000, from which she furnished her own people and exported largely. This large value we find derived largely from the farms. With such figures before us, a growing population, and a surety that as cost of production is decreased by skillful management, the consumption of poultry products will be largely increased, we may rest assured of a market for some time to come. Now let us look at some of the reasons for making poultry culture a prominent department on the general farm, and especially on our New England farms.

First in importance is the small amount of capital necessary to invest. You have doubtless read Fannie Fern's story of the shrewd Yankee, who, wishing to start in the poultry business, borrowed from one neighbor a broody hen and from another a sitting of eggs. He soon had a fine litter of chicks and was ready to return the hen to her owner. But how was he to repay the eggs? He soon solved that by keeping the hen until she laid the required number of eggs, returned both hen and eggs, and "guessed he had as fine a litter of chicks as any one, and about as cheap, too."

Next is quick returns. One reason why a farmer cannot make money so rapidly as one can in many other lines of business is because he cannot turn his money over quickly enough. Poultry keeping will help the farmer in this respect by giving him steady cash returns if the business be rightly managed.

GREAT PROFITS IN PROPORTION TO INVESTMENT

Another and very important reason is greater profit. For the same investment of capital and labor no other department of the farm will yield such generous returns. Dollars and cents are what all of us are striving for in business. So this is a most potent argument in its favor. You remember the old saying, "Take care of the cents and the dollars will take care of themselves." This is a most excellent piece of advice, but I think it would be still more applicable to the poultry business if it read like this, "Look out for the sense and the dollars will look out for themselves, ' for in no kind of work is good, plain common sense more valuable than in poultry culture. Another reason especially applicable to our farms that are at a distance from market is that it is a concentrated product, easy to handle and market at a distance, which is not true of more bulky products.

Still another reason is that waste products of other departments may many times be utilized, and instead of being a waste become a source of profit. For instance, dairying and poultry culture go hand in hand. When butter is made or cream sold, leaving the skim-milk at home, the milk will give far better results financially fed to poultry than when given entirely to swine, as is so commonly done.

Fruit and poultry make a good combination. The fowls aid you in the fight against insect pests and also much of what would otherwise be wasted is made to be of value.

Another point in favor of this industry is that you are continually enriching your farm and at the same time deriving

DEMAND AND SUPPLY

a profit from the business. This is an important point, for much of our New England soil has been managed in such a way in the past that the farmer of to-day has the difficult problem to solve of making a living and at the same time of bringing the soil from its worn-out condition to one of fertility. I have seen this done by means of poultry culture. A friend of mine has more than doubled the crop capacity of his land and almost entirely by this means. While I would not advise every farmer to take up poultry culture to the exclusion of other lines of farm work, yet it seems to me that there is a chance to make this a paying department on nearly every farm. Good markets are assured us in the many manufacturing towns and villages of New England for fresh fruits, vegetables, poultry and eggs, and as we can in no way compete with the western farmers in the cereals and many other farm products, it seems to me that the salvation of the New England farmer lies in producing the best and freshest of such products as our city and village people are and always will be so glad to obtain.

METHOD AND PATIENCE BRING SUCCESS

Now for a few thoughts as to how this line of work can be made a practical success. First, there are personal traits of character which underlie success in any business, and these must naturally be possessed or else acquired before we can look for the best results from a man's labors. He must have application, patience, persistence, and in every sense of the word be a hustler. Be on the alert for every new idea in your business, but do not be greedy and attempt to swallow more than you can digest. Season your scientific knowledge with lots of common sense, and, above all, run your business on sound business principles. Be a genuine Yankee, but do not "guess;" always know your business. Keep strict accounts and records, and study them. A good system of accounts is the surest guide you can have to success in any business, and you will find farming to be no exception, though comparatively few farmers keep them. Study your market, the particular likes and dislikes of your customers. Learn to fill every want, and just as they wish it, and never know more than your customers. If you wish to made changes in any way, do it in such a manner that they will think they are the ones making the change rather than you.

PROCEED WITH DUE CAUTION

Do not begin too expensively. Remember that every dollar you put into business is an interest-bearing factor, and must be accounted for out of your profits. Expensive or fancy buildings are not a necessity, but convenience of labor and proper conditions are. Make your plant cost as little as possible, but do not sacrifice convenience or proper conditions under any circumstances. Above all, look after the details, for no department of the farm needs so close attention to the many little details, or will suffer so quickly for lack of attention, as this. Careful attention to these littles, a love for the work, and a never failing will to succeed under any and every condition, will bring you success. Never depend upon luck, but always spell it with a "p," and never expect success till you have earned it.

THE EASTERN EGG MARKET

STRICTLY FRESH EGGS ALWAYS IN GOOD DEMAND IN NEW YORK, PHILADELPHIA AND BOSTON—WHERE THE EGGS COME FROM—HOW THEY ARE GRADED—SUMMARY OF RECEIPTS FOR NINETEEN HUNDRED AND FIVE—BOSTON PAYS HIGHEST PRICES

P. T. WOODS, M. D.

PRODUCING fancy fresh eggs for market is one of the most profitable branches of the poultry industry. Although egg farms are abundant throughout the East particularly in the neighborhood of New York, Philadelphia and Boston, their output has never been sufficient to meet the demand for strictly fresh eggs, and there is no indication at the present time that there is ever liable to be any danger of overproduction. The demand for high grade eggs is steadily increasing year after year and although receipts have shown slight gains there has never been a sufficient supply of high grade, fancy, fresh eggs in winter to appreciably lower the price.

No other branch of the poultry business offers such opportunities for making fair and certain profits in return for honest, sensible effort. There are many men in the East who started in a small way and have in the course of a few years built up egg farms that are now paying them comfortable incomes. There are very few farms that are not suitable for the development of an egg-producing business, and the matter of location is of less importance than a disposition to persevere in the undertaking and to hustle for business on the part of the man at the head. With the transportation facilities which now make the city markets so accessible to the country towns and villages there are very few places where eggs cannot be profitably produced and shipped to the large markets, even when at considerable distances from the producer.

Eggs can be more easily graded, packed and shipped than almost any other farm product. To secure a good trade the chief requisites are to establish a reputation for producing first quality eggs that are strictly fresh and to be able to "deliver the goods." The man who only makes occasional shipments and then of indifferent quality or poorly graded will never make a success of the egg business.

If the producer will take pains to establish a reputation for quality and maintain it with high grade goods he can rest assured that the consumer will attend to creating the demand for them. A well-known market poultryman when writing on this subject once aptly said:

"There is no more fussy individual than the consumer of table eggs, consequently when you have given him what he likes and wants, you are assured of a steady customer and —though there may now and then be some grumbling—fair prices. Eggs shipped from a distance, which may be relied upon to be true to the mark in color and quality, will find a ready sale in average times and conditions of the market, and are always sought after by the better class of trade.

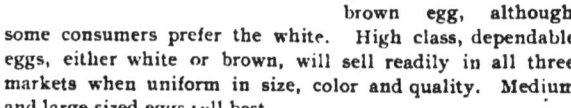
21—YOUNG FAVEROLLE BREEDERS

FANCY FRESH EGGS—WHAT THEY ARE

While the demand for strictly fresh fancy eggs is always in excess of the supply the producer has first to prove the quality of his goods before he can expect to succeed in this market. It is not sufficient to send in one or an occasional lot of good eggs, and the producer's word does not always pass as valid guarantee of quality. The goods must prove up in the market and the producer must prove his ability to provide a regular supply of equally good quality. Once a reputation for high class goods is established, it is comparatively plain sailing, provided care is taken to see that the quality is always strictly up to standard. Fancy fresh eggs must all be strictly fresh, but this alone is not sufficient. They must run uniform in size, color and condition as well as in quality of contents. The color most in demand depends somewhat upon the market, and lots of uniform color sell best. New York City prefers, and often pays a premium for a fine, white-shelled egg, while Boston calls for a rich brown color. Philadelphia leans toward the brown egg, although some consumers prefer the white. High class, dependable eggs, either white or brown, will sell readily in all three markets when uniform in size, color and quality. Medium and large-sized eggs sell best.

The quality of the fancy egg must be beyond reproach. The shell must be well filled and have a very small air space to prove its freshness. The contents must be firm and not watery or flabby. Thin watery albumen or pale flabby yolks when the egg is broken by the consumer means loss of custom and such eggs cannot be classed as "fancy." Watery eggs are most commonly produced by birds that are out of condition or that are not properly fed. Blood clots in the eggs or strong flavors are not tolerated in the fancy article. To avoid blood clots candle your eggs before you ship them. Strong flavors can be prevented by feeding properly, avoiding tainted meat, fish, swill, onions, poor, ill-smelling scraps and other foods that "taste" the eggs,—by using clean sweet cases and keeping your egg room as fresh and clean as scrubbing and fresh air can make it. You may catch a buyer once with inferior goods but seldom twice.

WHERE THE EGGS COME FROM

While a portion of the eggs in our big eastern city markets are from eastern egg farms and known as "nearby fancy hennery" stock, the supply is small when compared with the quantities of eggs received daily from West, South and "Way Down East." Many of the nearby egg farmers cater to a special private wholesale or retail trade, and such do not appear in the market receipts. Many thousands of dozens of such eggs are marketed monthly in the smaller towns and cities as well as in the larger ones, and of this trade it is impossible to obtain a record or to even fairly estimate it. Eggs are shipped from many states to these large city markets. Some of the best average eggs in size and color are received from the West, and of late the South has been supplying fresh eggs of excellent quality that fetch good prices. As a rule, however, southern eggs run small and are of inferior quality.

All eggs are now packed in regular egg cases containing thirty or thirty-six dozen. These cases are light weight, clean wooden boxes divided into two compartments by a wood partition. Each side will hold from fifteen to eighteen dozen and is fitted with pasteboard fillers for dividing the space into smaller compartments for each egg, three dozen to a layer on each side of the case.

A regular shipper of nearby fancy fresh eggs usually has his own private cases bearing his name and address, and marked to be returned when empty. Such cases are of heavier build to stand frequent shipment, and are usually made of a good grade of soft pine lumber.

This class of stock should be packed and shipped separately." Wholesalers open all packages of eggs received, candle and sort them before they are turned over to the wholesale or jobbing trade. This is the invariable rule except where cases are sold in large lots subject to the marks and grades of wholesalers.

Candling the eggs means that they are tested by passing them before a strong light so that the contents may be judged and the eggs properly graded. The candler notes the size of the air space, the appearance of the contents and quickly detects any stale, bad, bloody or otherwise imperfect eggs. By testing the eggs in this manner they are sorted into regular market grades that can be sold with a guarantee as to the quality that may be expected. Candlers become very expert and the average day's work for one man is estimated at from fifteen to twenty cases in ten hours. In some large cold storage houses special testing equipment is employed and here the eggs can be candled and graded much more quickly.

The Boston Fruit and Produce Exchange publishes the following rules on the classification and grading of eggs. Similar rules are in force in all large markets.

22—THREE VIEWS OF AN EXCELLENT COLONY COOP

Some of the best western eggs come from Indiana, Michigan and Ohio, while in the early spring months eggs of fairly good quality come from far away Missouri. The better quality nearby eggs come from all sections of New England, New York State, eastern Pennsylvania, New Jersey and northeastern Maryland.

New York City draws largely on Rhode Island, Connecticut, New Jersey, eastern Pennsylvania and eastern New York for the bulk of its nearby eggs. The best southern eggs come from Kentucky, and the lesser grades from Tennessee and the South.

Philadelphia gets its nearby eggs largely from eastern Pennsylvania, New Jersey and Maryland.

Boston markets' nearby stocks are drawn largely from eastern Massachusetts, Rhode Island, Connecticut, Maine, Vermont and New Hampshire. The chief source of western eggs in all three markets has already been given.

HOW EGGS ARE GRADED

The importance of sorting eggs as to size, color and quality and packing each grade separately for shipment is evidenced by the fact that nearly all dealers print conspicuously in their trade-circulars the following instructions:

"Shippers should bear in mind that to command the outside prices, all poor, dirty and small eggs should be taken out.

"SECTION 1. Eggs shall be classified as Fresh and Storage. Eggs shall be graded as: Fresh—Extras, Firsts, Seconds, Dirties and Checks. Storage—Extras, Firsts, Seconds and dirties.

SEC. 2. Loss-off, as used in these rules, shall comprise all rotten, broken, leaking, heavy spots, broken yolked, hatched, blood veined and sour eggs. Very small, very dirty, light spots, cracked (not leaking), badly heated, frozen and shrunken eggs shall be counted a half loss in all grades except Dirties and Checks.

"Fresh Gathered Extras shall be free from small, dirty, cracked, heated or frozen eggs, and shall contain perfectly fresh, full, strong, sweet eggs, as follows:

From Feb. 15 to May 15............90 per cent.
" May 15 to Oct. 31............80 " "
" Nov. 1 to Dec. 31............70 " "
" Jan. 1 to Feb. 15............80 " "

"The balance, other than the loss, may be defective in strength or fullness, but must be sweet. There may be an average loss of one dozen per case, but if the loss exceeds this by not more than 50 per cent the eggs shall be a good delivery upon the allowance of the excess.

"Fresh Gathered Firsts shall be reasonably clean and of

good average size, free from frost, and shall contain fresh, reasonably full, strong, sweet eggs, as follows:

From Feb. 15 to May 15..............85 per cent.
" May 15 to Oct. 31..............65 " "
" Nov. 1 to Dec. 31..............50 " "
" Jan. 1 to Feb. 15..............65 " "

"The balance, other than the loss, may be defective in

23—A CONTINUOUS LAYING HOUSE

strength or fullness, but must be sweet. There may be an average loss of one dozen per case, but if the loss exceeds this by not more than 50 per cent the eggs shall be a good delivery upon the allowance of the excess.

"Fresh Gathered Seconds shall be reasonably clean and of fair average size, and shall contain fresh, reasonably full, sweet eggs, as follows:

From Feb. 15 to May 15..............70 per cent.
" May 15 to Oct. 31..............40 " "
" Nov. 1 to Dec. 31..............30 " "
" Jan. 1 to Feb. 15..............10 " "

"The balance, other than the loss, may be defective in strength or fullness, but must be merchantable stock. There may be an average loss of two dozen per case. If the loss does not exceed this by over 50 per cent the eggs shall be a good delivery on allowance of the excess.

"Fresh Gathered Dirties shall be sweet and free from cracks. There may be a loss of two dozen per case. If the loss does not exceed this by over 50 per cent the eggs shall be a good delivery upon allowance of the excess.

"Checks shall be sweet. Blind checks and cracked eggs (not leaking). There may be a loss of two dozen worthless egg per case. If the loss does not exceed this by over 50 per cent the egg shall be a good delivery upon allowance of the excess."

December 30th, the following change in the official grading of eggs went into effect in the New York market:

Extras, 80 per cent. fresh, 1½ dozen maximum loss.
Firsts, 65 per cent. fresh, 2 dozen maximum loss.
Seconds, 50 per cent. fresh, 3 dozen maximum loss.
Thirds, 30 per cent. fresh, 6 dozen maximum loss.

The grade of "extra firsts" formerly used is eliminated.

HOW EGGS ARE MARKETED

The large western wholesale houses supply practically all of the western eggs to eastern markets either by direct agents or through brokers who buy and sell on margins of so much per dozen. Even the nearby eggs are as a rule supplied by dealers or regular shippers, and comparatively few small lots are handled by the regular commission houses. The commission men and jobbers prefer to deal with established, responsible and easily located firms or shippers who can be quickly found in case of trouble or disputes as to quality and adjustments promptly made. With small local shippers the additional trouble and expense in finding the shipper and adjusting differences frequently make the transaction too burdensome.

As a rule the egg trade is confined to the city and to the towns and cities in the immediate vicinity, the greater portion of the trade being within the city itself. Nearly all of the outside cities and towns have local firms who do the bulk of the trading in eggs. Every county and most large towns in localities where eggs are produced have their own local egg buyer who makes a business of collecting eggs, paying the producer a slight reduction from the prevailing wholesale price. He in turn sells to the wholesaler or retail man and in some cases may also supply a private egg trade.

Egg farmers who are in the business on an extensive scale often make a practice of buying up the fresh eggs from the lesser plants in their vicinity and so take care of the shipments to the city, practically controlling the trade in their neighborhood. In all towns of any considerably size egg dealers are equipped with cold storage facilities so that the local dealers have the same advantages for handling and keeping eggs as the large city dealers.

Strictly fresh fancy eggs are usually largely sold to private trade or to the better class of retail stores in towns or cities near the plant producing them. However, all of the three large markets mentioned herein receive considerable shipments of these strictly fancy eggs and as a rule they bring a premium over the prevailing market prices.

Boston has earned the reputation of having the most fastidious trade in fancy eggs and the best quality goods frequently bring an appreciable advance over the highest

24—INCUBATOR CELLAR AND NURSERY

quotations. Undoubtedly this market pays the highest prices in the United States for strictly fancy, fresh, brown eggs during the winter season. This winter having been an open one the prices have not soared as high as during the past two winters, when eggs reached 75 cents per dozen at retail, but even this season retail sales have been reported during late November and early December at 50 and 60 cents per dozen for especially fancy quality.

DEMAND AND SUPPLY

EGG RECEIPTS AND PRICES

The magnitude of the egg business in our large eastern cities will be better appreciated after consulting the table published herewith showing the "Eastern Egg Market Summary for 1905." This shows the weekly receipts of eggs in cases in New York City, Philadelphia and Boston for each week during the year. I am indebted to the weekly market letters in the "New York Produce Review and American Creamery" for the weekly receipts, from which the totals and estimates are figured. These figures do not tell the whole story since there are thousands of cases sold to private, local and special trade that are not included in these regular market reports. In figuring the dozens of eggs received I have used the 30-dozen case as the unit, although this is undoubtedly under-estimating the receipts as eggs are shipped in both 30-dozen and 36-dozen cases.

A table showing the highest wholesale price per dozen quoted each week in the year 1905 for strictly fresh eggs is also given for all three cities and will prove interesting for comparison, both as to prices and with the table of receipts. These prices are from the regular weekly market review and may be taken as conservative. Undoubtedly in many cases some lots of eggs actually sold higher than the prices quoted,—strictly fancy nearby eggs sometimes wholesaling at from two to five cents more per dozen. It will be noted that the average highest price per dozen for the year in Boston was 29 cents, while New York City was 24 and Philadelphia 23½.

Taking the egg trade as a whole during the past two or three years the general average price of fresh eggs for all grades in New York and Philadelphia has been 20 cents per dozen, while Boston has run about five cents per dozen higher. In making the estimate of total values of the 1905 receipts these latter average prices were used.

As a sample of the variation of the price in the different grades of eggs in Boston market the following is quoted from the weekly review for the week ending August 12th., when it will be noted the highest price in the table is quoted as 27 cents per dozen.

EGGS—QUOTATIONS AT MARK

Fancy Hennery	26 to 27
Maine, Vermont, New Hampshire and York State Extras	24 to 25
Maine, Vermont, New Hampshire and York State Fair to Good	21 to 22
York State Firsts	22
Michigan Extras	22
Michigan Firsts	20 to 21
Western Firsts (northerly sections)	19½ to 20
Other Western Firsts	19 to 19¼
Western Good to Fair	17½ to 18
Western Poor to Fair	15 to 16
Western Dirties	11 to 14
Western Checks	10 to 12

There is one thing the beginner in egg farming needs to learn and should print large and paste in his hat, and that is "Aim to produce eggs that can be sold at a profit at the lowest market prices." It is possible in nearly all sections of the country to produce egg that can be sold at a small profit at from 10 to 12 cents per dozen. The man who can accomplish this is sure of good profits from the entire seasons production. It has been done, is being done at the present time by practical men who make their living from poultry, and it can be done again by you or any other beginner who has plenty of grit, push and "git thar" ability in him.

WHERE THE EGGS GO

Only a comparatively small proportion of the eggs received are sold for export, the supply of strictly first-class eggs frequently being short of the demand for home consumption. Of course the great bulk of eggs are disposed of for culinary purposes, either in the home, hotels, restaurants, ocean steamships, bakeries, etc. The greater portion of strictly fresh fancy eggs are disposed of to high-class family trade or to the better grade of saloons, drug stores, etc., where fancy drinks are sold. The number of eggs used in the preparation of fancy drinks in our large cities would figure up to an immense total were it possible to get the exact figures.

As a rule the hotels, restaurants, etc., depend largely upon western eggs, particularly in cool weather, buying the nearby eggs only when warm weather lowers the prices and when the eggs shipped from a distance are not as dependable. The better grades of western eggs are of good color, size, quality, well filled, and usually answer the purpose admirably. When summer weather forces this trade to fall back on the nearby eggs this increased demand materially aids in keeping up the price for locally produced stock. Bakeries and the cheaper

EASTERN MARKET EGG SUMMARY FOR 1905

FOR WEEK ENDING	CASES RECEIVED WEEKLY N. Y.	PHILA.	BOSTON	Highest weekly wholesale prices per doz. for strictly fresh eggs. N. Y.	PHILA.	B'ST'N
Jan. 7	29,349	13,684	9,379	$ .29	$ .32	$ .35
Jan. 14	35,403	15,339	15,134	.31	.28	.35
Jan. 21	46,769	9,578	14,027	.28	.25	.33
Jan. 28	33,200	6,580	9,022	.30	.29	.34
Feb. 4	39,448	10,256	11,738	.32	.30	.35
Feb. 11	21,374	6,390	8,068	.33	.30	.32
Feb. 18	13,935	5,435	5,618	.35	.35	.35
Feb. 25	18,901	8,282	4,469	.30	.30	.34
Mar. 4	18,883	10,539	4,124	.35	.31	.35
Mar. 11	42,834	14,152	12,432	.28	.28	.29
Mar. 18	111,945	18,452	29,188	.18	.17½	.20
Mar. 25	142,633	24,287	45,711	.17½	.17½	.21
Apr. 1	119,319	26,890	53,252	.17½	.17	.21
Apr. 8	149,331	25,361	67,640	.19	.18	.21
Apr. 15	142,188	29,408	70,937	.18½	.17½	.20
Apr. 22	144,720	30,142	81,040	.18	.17	.20
Apr. 29	129,965	25,998	66,175	.18	.17½	.20
May 6	123,056	20,583	58,144	.18½	.18½	.20
May 13	122,606	28,268	45,666	.19	.17½	.21
May 20	132,166	27,983	41,614	.18½	.18	.21
May 27	124,965	19,065	37,855	.17½	.17½	.20
June 3	102,967	21,579	38,112	.18	.18	.21
June 10	106,054	34,401	35,732	.17½	.17½	.21
June 17	105,720	30,926	26,441	.17½	.17½	.22
June 24	89,154	17,971	28,843	.17	.17½	.22
July 1	79,922	16,269	21,471	.17½	.17½	.22
July 8	71,469	21,851	18,406	.17	.17½	.23
July 15	70,730	18,164	21,929	.19½	.18	.25
July 22	81,704	18,959	19,485	.19½	.18½	.26
July 29	72,187	14,683	22,146	.19½	.18½	.27
Aug. 5	65,720	14,947	24,125	.21	.19½	.27
Aug. 12	47,563	13,965	29,567	.21	.21	.27
Aug. 19	55,988	13,827	34,257	.23	.22	.28
Aug. 26	67,669	16,696	29,508	.22	.22	.28
Sept. 2	70,242	15,301	26,738	.21	.24	.28
Sept. 9	63,992	22,611	22,864	.22	.23	.29
Sept. 16	57,288	16,468	24,922	.22	.22	.30
Sept. 23	58,360	16,414	23,510	.22	.22½	.31
Sept. 30	62,996	16,018	26,965	.23	.23	.32
Oct. 7	60,332	18,912	32,414	.23	.23	.32
Oct. 14	55,623	16,415	28,596	.23½	.23	.32
Oct. 21	42,472	15,002	22,280	.25	.26	.34
Oct. 28	33,893	12,171	17,610	.27	.27	.37
Nov. 4	42,114	13,808	14,881	.27	.29	.38
Nov. 11	43,648	15,941	16,060	.28	.29	.40
Nov. 18	33,667	12,283	10,878	.31	.31	.42
Nov. 25	39,022	11,495	16,756	.32	.32	.40
Dec. 2	33,049	8,536	10,249	.32	.32	.42
Dec. 9	35,287	8,477	11,944	.29	.29	.42
Dec. 16	34,224	12,596	15,396	.28	.29	.40
Dec. 23	26,518	7,767	12,939	.30	.29	.40
Dec. 30	31,173	10,080	14,058	.29	.28	.35
Total cases rec'd and av'ge price	3,581,627	886,705	1,393,456	$ .24	$ .23½	$ .29

	N. Y.	PHILA.	BOSTON
Total Cases Received 1904	3,363,630	1,011,164	1,122,819
Dozens at 30 per Case 1905	107,448,810	26,591,150	41,803,680
Estimated Total Value 1905	$21,489,762	$5,318,230	$10,450,920

Note.—The estimated average price of eggs in New York and Philadelphia markets is usually set at 20 cents per dozen, while that of Boston market is 25 cents per dozen. In basing estimated value in this table we have used these figures as the average prices. As eggs are shipped in 30 and 36 dozen cases the above estimate based on 30 dozen cases may be considered a conservative one.

Philadelphia receipts for 1904 were partly estimated, owing to a change in the method of reporting receipts, and it is probable that they are slightly over-stated, so that the decrease in receipts for 1905 is really not as great as it appears to be.

restaurants use the cheaper and poorer grades of eggs. Besides the large quantities of eggs used in this manner for feeding the people, great numbers are also used in the different arts and mechanical pursuits. Quantities are used by manufacturers of fine leather, photographers supplies, the preparation of mucilage and many other technical uses.

There is always a fairly good outlet for all checked or cracked eggs, dirty eggs, and even eggs that a fastidious person would consider wholly bad. As a rule, cracked, broken and "heavy" eggs, (those with unsettled and displaced yolks,) and such as are quite stale yet not wholly bad, are broken and the contents emptied into cheap tin pails, either entire or the yolks and whites separated. These pails when full are covered and put into freezers where they are frozen solid and kept so until they are needed for use. This supply of canned eggs is drawn on by the bakers and confectioners for the manufacture of cakes, pies, candies and other edibles. Many of these manufacturers of delicacies find it more convenient

A great deal of attention has been given by experimenters in poultry work to the means whereby winter eggs can be produced in profitable numbers. It has been demonstrated by practical tests that it is desirable to breed for egg production as carefully as we would for other practical or standard qualities. Feeding is probably of less importance provided the birds are always supplied with pure water, a good variety of wholesome grain, meat and green food.

The best results in winter egg production are undoubtedly obtained with early hatched pullets, since old hens recuperating from their fall moult seldom produce, with any regularity, a good egg yield before late winter and the early spring months. Frequently the old hens do not really get down to business until March so that it is absolutely necessary that we have pullets, particularly early hatched pullets upon which to depend for the winter egg production if the best prices are to be obtained.

It is commonly stated that pullets will come to laying

25—COLONY HOUSES SUITABLE FOR SMALL CHICKS

and decidedly cheaper to buy eggs of this sort, particularly when the whites and yolks can be purchased separately.

It has been stated that these frozen eggs cannot be successfully used in making custards until they are first put through a fine sieve which breaks up the lumps and renders the mass smooth. This should prove an invaluable hint to the house-keeper who want to use home grown, damaged or frozen eggs in the manufacture of custards, custard pies and other delicacies. Even the very bad and absolutely rotten eggs find a market, a considerable portion of them being saved. According to reliable data it has been estimated that many of the larger dealers in eggs are able to get from $25 to $50 a year for absolutely rotten eggs. These eggs are put through a special clarifying process and the good parts separated, and these ultimately in the hands of the baker are turned into various edibles that are said to delight the hungry stomach. Such parts as are too far gone to give good results in the clarifying process are largely utilized in the manufacture of mucilage, and to some extent in other branches of the different arts and mechanical pursuits.

WINTER EGGS PAY BIGGEST PROFITS

By referring to the tables it will be noted that the time of best prices for fresh eggs is from the first of September until about the middle of March, while the lowest prices are usually reached during April and May. These facts make it evident that the best profits will be obtained by those able to produce economically the greatest number of winter eggs.

maturity in five or six months, but the egg farmer cannot afford to rely upon this rule. Seven months old at laying maturity for the American varities is probably nearer the actual average.

The writer prefers to get out his early pullets between the first of March and the middle of April, sometimes a week or so earlier but seldom later. Wyandotte pullets hatched at this time of the year have always given us the most satisfactory results, and thirty-eight of them at the present writing have been giving us since before Thanksgiving time a regular daily egg yield of from 16 to 25 eggs. With the heavier American variety the Plymouth Rocks, it will be wiser to get them out during the month of March and not later than the first week in April, since they are somewhat slower in coming to laying maturity than the Wyandottes and Rhode Island Reds. Some of my best layers this winter were hatched on the 22d of February and the 4th of March, and I believe that as a rule egg farmers will find, regardless of variety, that the March hatched pullets prove the most desirable winter egg producers.

To obtain these early pullets it is of course necessary to rely upon the modern incubator and brooder, since hens cannot always be depended upon for hatching and rearing so early in the season. This brings us to a discussion of artificial incubation and brooding which does not lie within the province of this article, except that we may state that the modern incubator and brooder has made it possible for poultrymen to produce early pullets in sufficient numbers to successfully cater to the winter fresh egg trade.

THE WESTERN EGG MARKET

WHAT CHICAGO, KANSAS CITY AND SAN FRANCISCO WHOLESALERS PAY FOR EGGS—RETAIL PRICES WILL RANGE FROM ONE TO TEN CENTS MORE—AN EVIDENT UPWARD TREND IN THE PRICE OF EGGS—SUPPLY DOES NOT EQUAL THE DEMAND

THE EDITOR

Western poultrymen will view with great satisfaction the average prices paid for eggs, as shown in the following tables, when they compare these prices with the average price of eggs in the great eastern markets of New York, Philadelphia and Boston, as given by Dr. Woods on page 29. Heretofore western egg producers have sighed for the high prices paid by eastern markets. Time was when in the Mississippi Valley eggs commonly sold in early summer for twelve and fifteen cents a dozen. But those days are happily gone—forever, we hope. With the average price for 1905 on the western coast a fraction of a cent higher than the high-water average price of Boston, and with Chicago within three and one-half cents of the average New York prices, western producers may sit up and begin to take notes. Let "improve the quality and then increase the quantity" be the watchword and "western" eggs will cease to be a drug on eastern markets and they will command top prices at home and abroad.—Editor.

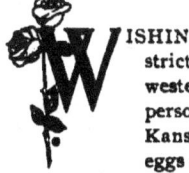

WISHING to compare the prices received for strictly fresh eggs in different parts of our western country we wrote to a responsible person in each of the three cities, Chicago, Kansas City and San Francisco, for prices of eggs during the past year. Following are the three reports:

REPORT FROM CHICAGO

CHICAGO, ILL., January 5, 1906.

"I am pleased to give you an official report which was obtained from Mr. A. W. Hale, Secretary of the Chicago Butter and Egg Board."

EGG QUOTATIONS, CHICAGO

1904		1905	
January	26-33	January	24-31
February	18½-32	February	29-34
March	15½-18	March	15-29
April	16-17½	April	16-18
May	15½-16½	May	16-17
June	15½-16	June	15-15½
July	16-18	July	15½-19
August	28-18½	August	18½-20
September	18½-20	September	19½-20½
October	20-22	October	20½-23
November	22-26½	November	23-28
December	23-28	December	24-29
Local receipts, 2,232,768 cases.		Local receipts, 2,445,532 cases.	

REPORT FROM KANSAS CITY

KANSAS CITY, KAN., January 3, 1906.

"As requested, I have procured what I think are the figures that you wish, showing the average price of eggs on the Produce Exchange in Kansas City for each month of the past year. The prices averaged as follows:

1905

January	.23
February	.28
March	.18
April	.15
May	.15
June	.13
July	.14
August	.17
September	.17
October	.18
November	.23
December	.25

I tried to get the figures for the total receipts for each month, but the Secretary did not have them in shape, or so compiled as to give them to me. The demand for eggs here is equal to the supply, no matter how large the supply may be. Buyers are here from all parts of the east, buying for their home markets, as well as for cold storage purposes, as soon as eggs become plentiful and cheap enough to warrant their use for storage. During these months New York and Boston dealers keep a buyer in the Produce Exchange here, and a great many car loads are shipped from here to the east through this channel. The eggs are sold outright and not on commission. The packing houses are heavy buyers. Armour and Swift handle a large per cent of all the eggs coming into this market, which amount to a great many hundred car loads a year."

REPORT FROM SAN FRANCISCO

SAN FRANCISCO., CAL., January 5, 1906.

"The main supply of eggs for the San Francisco market is shipped in from the counties of Sonoma, Napa and Santa Clara. So-called "store eggs" are received from the states of Kansas, Oklahoma and Missouri.

Importations into the state of California from these three states for the year commencing April 1, 1904, and ending April 1, 1905, amounted to 2,964,000 dozen. There was also a considerable quantity imported from Japan. Following are the wholesale prices of eggs here, from January 14 to December 29, 1905:

EGG QUOTATIONS, SAN FRANCISCO

DATE—1905	RANCH	CHOICE	GATHERED	STORE
January 14	30-32	27½-29	28	25
January 28	27-28	26-27	25	23
February 3	27-28	26-27	25	23
February 18	23-24	22-22½	22	21
February 24	21	19-20	20	19
March 2	16	15	14	14
March 10	20	18-19	16	14
April 7	18	16-17	16	14
April 21	19	17-18	17	14
April 28	18	17	17	14
May 5	18	17	17	14
May 12	18	17	17	14
May 18	19	17	17	14
May 25	20	18	18	16
June 2	20	18	18	16
June 9	22½	20	19-20	18
June 16	20	18	18	17
June 23	20	18	19	18½
June 30	20	18	18-19	16-17½
July 7	20	17-19	17-17½	15½-17
July 14	23	19-21	17-17½	15-17
July 28	26	22½-25	20-22	17
August 4	29	27-28	25-26	16-22
August 11	30	28-29	27-28	16-22½
August 18	30	28-29		
August 25	30	28-29	27-28	16-22½
September 1	30	28-29	27-28	16-22½
September 8	32½	30-31	27-30	20-22½
September 15	35	30-34	27-30	20-25
September 22	37	32-36	29-32	24-28
September 29	39	35-38	32-35	26-30
October 6	38	35	30-35	26-28
October 13	40	37	35-37	26-30
October 20	43	37½-42	37-42	26-30
October 27	48-49	43-44	40	24-25
November 3	48-49	44-46	40	24-25
November 10	51	46-48	40	24-25
November 17	51	46-48	40	24-25
November 24	50	46-48	40	24-25
December 1	42½	39-40	28	23-24
December 8	46	35-40	29	23-24
December 15	51-52	30-45	29	23-25
December 22	30-31	26-30	27-28	23-24
December 29	37	28-34	30	21

EGGS AND EGG FARMS

GLEANINGS FROM FIGURES

Let us see what we can learn from these figures. We find that in Chicago the average prices paid in 1904, beginning with January, were 29½, 25¾, 16¾, 16¾, 16¼, 15¾, 17, 18¼, 19¼, 21, 24¼ and 25½ cents per dozen. For 1905 the average prices were 27¼, 31¼, 22, 17, 16¼, 15¼, 17¼, 19½, 20, 21¾, 25½ and 26¼ cents per dozen.

The average price per dozen for the six months of less production—from September 1 to February 28, inclusive—in 1904, was 24½ cents; in 1905 it was 25½ cents. For the six months of greater production—from March 1 to August 31, inclusive—in 1904, was 16¾ cents; in 1905 it was 17½ cents. Quite a difference between the prices of summer and winter eggs—over 7 cents.

The average price for the whole year 1904 was 20½; for 1905 it was 21¼. Only one cent difference, but it is on the right side of the ledger to please us.

In 1904 the highest price paid was in January, 33 cents; in 1905, 34 cents was paid in February. The retail price was probably 40 cents, if not higher, in these two months. It is reasonable to believe that the producer or the merchant who sells to the best custom could get 50 cents per dozen for guaranteed, new-laid winter eggs. But how many weeks or months would they be willing to give that much, is another question. It is clear that prices are gradually advancing.

How many hens produced the eggs furnished the Chicago market as shown by the market report? Let us see.

First, we note that the receipts were 212,764 more cases in 1905 than in 1904—a fair increase. A case usually holds thirty dozen. That means 73,365,860 dozens of eggs were recorded as received in Chicago. Assuming that hens generally produce eight dozen eggs per year (the government reports for 1900 recorded less than six dozen), we find it takes 9,170,745 hens to supply Chicago with eggs—fresh and otherwise. It does not look as though the market stood any chance of being over supplied, does it? Chicago is only one large city. It makes one dizzy to try to comprehend one flock of nine million—suppose they should cackle at once! Try to imagine the immense number of hens needed to produce the total egg crop, and one is bewildered and becomes almost incredulous.

KANSAS CITY PRICES LOWER

We find the prices in Kansas City are lower, probably because of that city's position near the country's greatest egg-producing states. And even though this is true, you will note our correspondent states that the demand always meets the supply. And you need not be afraid any other condition will prevail during your lifetime.

The average price at Kansas City for 1905 was 19 cents, or over two cents less than at Chicago. For the less productive months, i. e., fall and winter months, Kansas City paid 22⅔ cents as against Chicago's 25½ cents. For the spring and summer months, the figures are, Kansas City, 15½, Chicago 17½. We have only the average prices given here but it is fair to suppose that in the two months when the average price was 25 cents, the maximum wholesale price reached 30 cents or better. Undoubtedly the farmers find it profitable to furnish eggs to the Kansas City market at these prices.

TABLE OF AVERAGE PRICES OF EGGS

	SEPT-FEB. 1905	MARCH-AUG. 1905	FOR YEAR 1905
Chicago	25½	17½	21½
Kansas City	22⅔	15½	19
San Francisco—			
Ranch	36½	21½	29¼
Choice	33⅔	19⅔	26¼
Gathered	30⅖	18 4-5	29 3-5
Store	23½	15 8-9	19¾

Does not that 51 cents wholesale price paid for "ranch" eggs in San Francisco fairly make one's mouth water? Note that "store" eggs (which seem to be in poorest repute) brought as high as 30 cents at one time. And note also that eggs were imported from Japan. No wonder so many persons are tempted to leave more rigorous climates for one so mild, when every one thinks himself perfectly capable of raising unnumbered chickens which shall lay numberless eggs, all of which are to find ready sale at 51 cents per dozen! With a clean, smooth board and a stub of a pencil one can "figure" himself into a millionaire in about fifteen minutes.

No doubt there are serious drawbacks to producing "ranch" eggs when they are 51 cents a dozen, else the market would be overstocked, but at all seasons of the year they command the highest market price, showing that quality counts always. But some are produced, hence more can be. As in other parts of the country, when the right man (or woman) and the opportunity meet, good results are assured.

All these figures—all these words—resolve themselves into the following facts:

The demand exists. It exceeds the supply; hence there is always a market at remunerative prices.

New-laid, carefully graded eggs and prime market poultry (alive or dressed) find ready sale at a good profit to the producer. Therefore it behooves us to continue our crusade for "Better Poultry and More of It."

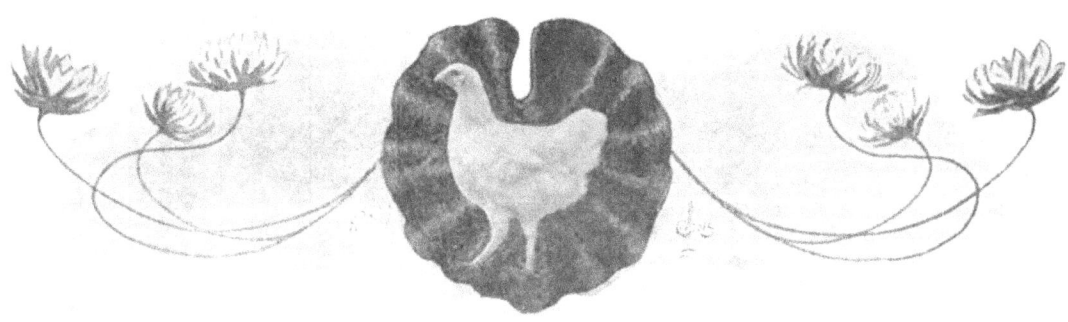

THE QUALITY OF EGGS

THE TECHNICAL TERMS USED BY COMMISSION MERCHANTS AND WHAT THEY MEAN—WHY PRICES OF EGGS VARY—SOME SUGGESTIONS TOWARDS IMPROVING THE QUALITY AND OBTAINING BETTER PRICES—STRIVE TO PRODUCE THE BEST

A. F. HUNTER

WE have been requested to explain the terms used in reports of the egg markets, such as appear in the following from "The Producers' Price Current."

N. Y. MERCANTILE EXCHANGE
QUOTATIONS OFFICIAL

QUOTATIONS FROM MARK—

Nearby, fresh gathered, extras	@20
Nearby, fresh gathered, seconds to firsts...15	@18
Western, fresh gathered, extras	@19
Western, fresh gathered, firsts...17	@18
Western, fresh gathered, seconds...15	@16
Western, fresh gathered, thirds...13	@14½
Western, fresh gathered, very inferior...10	@12
Western, fresh gathered, dirties, No. 1 (candled)...12	@13
Western, fresh gathered, dirties, No. 2...10	@11
Checked eggs	5 @ 9
Culls, per dozen	5 @ 8

NEW YORK DEMANDS WHITE EGGS

The term "nearby, fresh gathered, extras" is applied to the finest quality of eggs from nearby points, say from New England or the middle states. They would have to be eggs of good size, practically uniform in size, and fairly uniform in color. New York prefers white eggs, hence eggs of best quality for the New York market should be white; that does not mean "whitish," but "white," the chalk-white of Leghorn and Minorca eggs.

It is not necessary that eggs be marketed within one day, or two or three; they may be several days (or even a week) old and still have shrunk hardly perceptibly. If they have been kept in a cool place and the weather is cool they will be in the condition of "full, strong-bodied," which characterizes new laid eggs, and will bear the searching test of an electric light all right. That is, when held up to the testing light it will be seen that the shell is practically full, and when given a quick half-turn, a comparatively slow movement of the yolk indicates that the yolk and white are both of good consistency, not "fluid." In a stale egg both the increased size of the air cell and the fluidity of the body of the egg indicate to the practiced eye of the tester the quality (or grade) of the egg; if you will look at two or three eggs of different ages, say two, four and six weeks old, before a testing light, and then break them and pour the contents side by side on a plate, or sheet of white paper, you will note a difference in the way the eggs "stand up,"—the staler and more fluid the egg, the more it will "run," will spread out thin.

BROWN EGGS REQUIRED FOR BOSTON

Boston stickles for brown eggs, and to grade as the very best they must be of good size, fairly uniform in size and fairly uniform in color. They need not be a rich, dark brown, as some people think. There are varying shades of brown to be noticed in a case of good, brown, "nearby hennery" eggs;

the important points are that they shall be of good average size and good brown (and brownish) color. To get an idea of the average size of the finest eggs in Boston we went to Messrs. Chapin & Adams, bought three dozen eggs out of a case just in from a Maine farmer, weighed them, and found the weights were 25, 24½ and 25 ounces per dozen; a week or ten days later another lot of three dozen was bought, these, too, just shipped in from Maine, and they weighed 25, 25 and 25 ounces per dozen,—one dozen was a wee bit light and another good and strong weight, and the three dozen weighed all together gave a good four pounds eleven ounces, as against the first lot weighing four pounds ten and a half ounces. As these were

26—A TENT-SHAPED COLONY HOUSE

all eggs of year-old hens (it was in September), they probably were a trifle heavier than "nearby hennery" eggs will average, taking the whole year through, and we would say that eggs weighing a little better than a pound and a half to the dozen would be all right in size. To get further light upon this interesting question we went to another friend in the egg trade and asked him to sell us three dozen "fresh western" eggs, they weighed 4 pounds and 4 ounces; or 68 ounces, as against 74½ and 75 ounces, respectively, for the Maine eggs, a trifle more than 2 ounces difference in the dozen.

"QUOTATIONS AT MARK"

To return to the questions. "Quotations at mark" is a trade term meaning that the eggs are sold just as they are, without having been lighted or even opened, the buyer taking the risk of finding the eggs to average better or worse in quality than might be expected. For example, a car lot of western eggs from a known shipper may average three dozen (10 per cent) bad to the case; a buyer takes a quarter, or half, or all of them "at mark" at eighteen cents, when "fresh western" are quoted at twenty cents, thus throwing the ten per cent shrinkage off of the price instead of it being thrown out of the eggs themselves. Practically in buying "at mark" the buyer assumes the risk of shrinkage, whereas, if he bought at quotations he would pay for the grade of eggs bought and the western shipper has lost the shrinkage.

STRENGTH AND FULLNESS DETERMINE QUALITY

The "firsts," "seconds" and "thirds" are graded as to quaility, which, as explained above, is determined by the strength and fullness of the body of the egg; the weaker and more watery the egg and the larger the air space (which shows the amount of shrinkage that has taken place), the poorer in quality, and a "very inferior" egg would be pretty far gone, would be just upon the edge of the condition called rotten.

This also applies to "dirties No. 1" and "dirties No. 2"; the No. 2's are not any more dirty than the No. 1's, but are eggs of a grade poorer in quality. But why should people send "dirty" eggs to market? Why should they have "dirty" eggs at all? It is purely and simply "neglect," and any one having "dirty" eggs to market has only himself to thank for it! Those "dirty" eggs are (some of them) laid in stolen nests in the fence corners, or under a cluster of bushes, or under a broken down cart or old outbuilding, and may be a week or two weeks old and more or less soiled (stained) when found; never mind, into the basket they go, with the good eggs, and all are sold to the storekeeper in trade. Sometimes the eggs are laid at the base of a manure pile in the barn cellar, and the "stain" comes from contact with manure; of course the eggs themselves are somewhat "flavored" with the manure odor, never mind, put them into the basket and sell them in trade!

There is an outlet for dirty eggs, also for very "old" and partly decayed eggs, in the use that photographers, leather dressers, calico printers, etc., find for them; and there are people in the great cities so poor they are glad to buy such almost worthless eggs, and eat them for food!

DEMAND FOR CRACKED EGGS

A "checked" egg is one that is slightly cracked, but not cracked so much that the skin enclosing the egg proper is ruptured and the white is leaking out. There is quite a demand in our large cities for these slightly cracked eggs, and children are sent out from the tenements to visit the wholesale egg dealers in the hope of buying a dozen or two cracked eggs cheap. One dealer told us he sometimes made "checked" eggs out of those tested off as slightly off from the germ having started a bit. A little tap against the side of the case gave it a bit of a crack, then it was put in the lot of checked eggs.

27—INTERIOR VIEW OF TRAP-NEST

SELECT YOUR MARKET

You say, "Out here eggs are eggs, and all go for the same price." That is unfortunate because the man producing No. 1 eggs gets no more for them than Smith or Jones, who brings in No. 2 or No. 3 quality eggs. Why couldn't twenty or thirty farmers, who have good, pure-bred poultry, club together to ship their eggs to a New York or Boston commission dealer and get the better prices that would be paid for No. 1's? Do you recall the story about the town in Canada where a grain dealer had influenced his customers to keep "better poultry and more of it," with the result that the Toronto dealers are now inquiring for eggs from that town and paying a cent a dozen above the market price for them? Pure-bred fowls lay larger eggs (as a rule) and better colored; better looking eggs than the common barnyard scrubs kept on the majority of farms and the first step towards improvement is to get rid of the scrub hens and keep good pure-breds. Then, too, the average size of the eggs can be substantially improved by selecting the largest, best shaped and colored to hatch the chickens from. It is wonderful what an improvement can be made in two or three generations by this simple method, and we can have a new generation each year if we so elect; that means we can effect a revolution in our product in two or three years. We don't have to wait several years, as with the horse and cattle kind, to get results—we begin to get them six months after the chicks are hatched.

The "quality" of the egg itself, of the yolk and white, is dependent on the food ration of the fowls. If fowls are well fed, and have a "balanced" ration, they will lay eggs that very closely fill out the shells and that are "strong-bodied," are jelly-like (rather than watery) in consistency. Good, sound yellow corn gives the rich, yellow color of the yolk that is so much desired, and as egg yolks are rich in fats the fat of corn makes rich yolked eggs. Unfortunately corn is a very fattening food and tends to fatten the fowls unless much care is taken in the feeding of it and it is "balanced" by an ample allowance of green food, such as cabbage, turnips, etc., and the mash "extended" with cut and steamed clover or alfalfa. Clover has been most strongly recommended for a winter green food, but alfalfa is even better, its food value being nearly equal to that of wheat bran, and western farmers who have alfalfa at hand are fortunate indeed.

PRODUCE THE BEST QUALITY

In closing we want to urge our readers not to send second and third quality eggs to market, but to produce and sell only first quality. Dont keep "scrubs" and market "scrub" eggs; keep good pure-breds and market pure-bred eggs. The purebreds will lay better and pay better, their eggs will appear better in the market, the consuming public will appreciate them better and will call for more of them.

DEMAND AND SUPPLY

SHIPPING EGGS BY EXPRESS

HOW EXPRESS COMPANIES HANDLE EGGS—EGGS FOR HATCHING SHOULD BE SHIPPED IN BASKETS

An article written by an authority on express practices—and a poultryman. It explains the "habits" of the expressman; tells why he has to become an adept at sliding and throwing boxes, whether they contain eggs or iron, etc.; gives some "inside" information, and to avoid "controversy" the writer requests that his name and address be withheld.—Editor.

AS I AM an employe of one of the largest express companies in the United States, and acquainted with its various methods of handling freight, I wish to give my brother poultrymen a pointer on the shipment of fancy eggs. Last year I received two square boxes with small rope handles attached. The eggs were packed in fine shape; in fact, I believe they would have gone around the world with ordinary handling. The boxes were properly labeled, etc , but in spite of all this care one dozen very valuable eggs were smashed and and the interior of the boxes so permeated with egg yolk that one would imagine they had been used in a foot-ball game. There was one weak spot about those boxes—to my mind a very important defect—and that was the handles. If they had been equipped with a fair-sized wood or willow handle I would have had a fighting chance of getting my full quota of eggs unbroken.

Now for my reasons: An expressman is nearly always overworked, being obliged to transfer freight with lightning rapidity in order that the greedy express companies may cut expenses by employing the least possible help. If he doesn't dispose of about so much freight in a given time, woe to him; hence in transferring freight to and from trucks and to and from the ends of cars he becomes an adept at sliding and

29—ENTRANCE TO TRAP-NEST, DOOR CLOSED

easy, but the next box may be heavy and come down on the box of well packed eggs with a jam. It is only by this practice of sliding and throwing that an employe can meet what the express companies term "requirements of the service," so that the habit becomes fixed, and it is a lucky box of eggs that does not get the full benefit of the above service.

"HANDLE" FREIGHT

There is only one kind of freight that the expressman never throws, but hands from man to man, or tarries and sets it down, and that is freight with stiff, breakable handles that cannot be corded into a pile of miscellaneous boxes. It is the habit of messengers to reserve space in the car for such freight and a transfer man would never think of throwing handle freight, for it isn't the "habit." In the eyes of the service, boxes with small cord, rope or strap handles are not "handle" freight, as they may be corded in with the miscellaneous. The difference in the care of basket freight over that of box freight is as marked as is the difference in the service of a Pullman car and a buffer, as regards human freight. Do not take it for granted that I recommend baskets for fancy eggs. I prefer boxes with breakable handles. They take up more room, to be sure, and that isn't pleasing to the express company, but what is that to the poultryman—he pays the price.

MUST PAY LOSSES

It is the practice of express companies to "make a bluff" at not paying losses resulting from negligence or carelessness on their part, but poultrymen who suffer loss through them should stand pat and demand a settlement, dollar for dollar. When making shipments, report to the express agent the correct value of the goods, which will be the amount of money you have received for the particular goods, or are to receive from them. Have them put this amount on your receipt, and then save the receipt. If you state the correct amount and the goods are damaged or destroyed, you can collect one hundred cents on the dollar by proving the valuation. Express agents and the men back of them will tell you in stilted phrase that you have signed a "release," etc., agreeing therein to accept a trifling amount, but do not let them deceive you or dissuade you. They know the law and that they must settle. Insist on your rights and they will come to time sooner or later.

25—TRAP-NEST WITH DOOR AND NEST REMOVED

throwing boxes and having them come down right side up in their proper places, piling them up like magic, corded as nicely as you please. If a box of eggs comes along it may ride a little

EGGS FOR PROFIT

A WHITE LEGHORN EGG FARM

AN INTERESTING ACCOUNT OF A VISIT TO THE WHITE LEGHORN FARM OF C. H. WYCKOFF—NOT AN OUTSIDE DOLLAR INVESTED—FOWLS PAY FOR EVERYTHING— THE METHODS THAT MADE THESE RESULTF POSSIBLE—EGGS THE CHIEF PRODUCT

THE EDITOR

WHEN a man has succeeded to the extent that Mr. Wyckoff has, he owes it to his conscience to tell other people how he did it. We can afford to be our brother's keeper to that extent. It gives us pleasure to say that Mr. Wyckoff agreed fully with this view of the case and he told us how his success was attained. We desire in a plain way, in imitation of Mr. Wyckoff's modest recital, to tell our readers the story of this man's success with poultry. It is a story with few parallels, though there is nothing fabulous or improbable about it. It is a story of oneness-of-purpose, of steady effort, of intelligent application, of making the most of one's opportunities, be they great or small, which is the true measure of man or woman.

Years ago Mr. Wyckoff came to his present home a moneyless man. There lived near, on a sixty acre place, an old man who was very anxious to find a purchaser for his weed-grown farm. He wanted to sell so bad that a man without any money was able to buy. Mr. Wyckoff's father went security for the first payment (deferred) and loaned the son enough cash to half-way stock the farm, buy a few tools and put some seed in the ground. On entering upon the task of reclaiming and paying for this neglected farm, with its tumble-down buildings, Mr. Wyckoff had in mind the poultry business. He had touched the business lightly while living in Ohio, and was of the belief that there was good money in it for him. Being without funds he had to begin small. The first year he owned twenty-five mixed hens, which were housed in a ramshackle building. These he soon replaced with some Plymouth Rocks and Brown Leghorns. From eggs sold from these he saved up seventy-five dollars during one winter and spring and invested in White Leghorn eggs. A year later he had a flock of one hundred and eighty White Leghorn hens. This was his third year on the place. His hens now began to serve him in good earnest. It so happened that market eggs went up a kiting that winter, and in January alone his flock of Leghorns earned him ninety dollars in eggs laid. From that year on new buildings and parks were gradually added, the hens paying for everything as they went along.

Said Mr. Wyckoff: "It is the plain truth that after the money which I spent on those first twenty-five mixed hens I never invested an outside dollar in my poultry. They gradually earned every dollar—every cent that went to buy more hens, to build new houses and new parks. It took five or six years for me to get things fairly under way, but as I had no money I could not do otherwise. All I could rake and scrape out of the farm went for interest on the purchase price, or was put back onto the place in improvements of various kinds. To-day you see what I have as the result of fifteen years' work, with the help of my hens. The farm has been paid for these two years, and my hens did it. Yes, sir, they not only bought and paid for themselves and the quarters they live in, but paid every dollar of the purchase price of this farm. I am now selling about $4,000 worth of produce from this place each year, and fully $3,500 of it is in poultry and eggs. I keep cows and market butter, but my cows cost me all they earn and have for the past three or four years. During the past two years I have cleared $2,800 from my poultry."

THE HOUSES

Let us briefly describe Mr. Wyckoff's place as we found it. He has seven double poultry buildings 12 by 40 feet in size and one single house 12 by 20. The 12 by 40 foot houses are divided into two equal apartments, and each of the fifteen apartments opens into a park 33 by 128 feet. These parks are

EGGS FOR PROFIT

fenced with 1 by 3 inch unplaned pine pickets six feet in height. At the bottom laths are nailed to the stringer between these pickets to prevent the fowls picking each other through the fence. The houses are placed some thirty feet apart in order to make room for the width of the parks. The partitions in the houses are boarded up "so that the fowls in one side will not know what is going on in the other and throw themselves against the partition when I am feeding," explained Mr. Wyckoff.

The houses are built of common, unplaned, foot-wide barn-siding, with double roofs, shingled. The walls are of two thicknesses of the foot-wide boards, with building paper between, but no air space. No artificial heat is used. Last winter was a record-breaker for low temperature, The mercury went down to 36 below zero. In the above described houses the combs of a few male birds were slightly nipped, losing the slim points of some of the serrations, but Mr. Wyckoff reports that he has never yet found one of his White Leghorn hens with a nipped or frost-bitten comb. One important reason for this is to be found in the small windows in his houses.

Mr. Wyckoff uses board floors in his houses and told us

let down during severely cold nights. The top or cover of this inclosure can be made of burlap, or of thin boards, the latter preferred. By this arrangement, all the heat that is generated by the bodies of the fowls and that thrown off by their breathing will be confined in narrow limits, and will increase the temperature ten to fifteen degrees. The smaller the inclosure, the greater the warmth.

Hens yarded fifty to each park were kept by Mr. Wyckoff solely for market eggs, not for breeding purposes. Where hatchable eggs are wanted and the breeders have to be kept confined in moderate sized yards, one Leghorn male to fifteen to twenty females is the limit, where best results are sought. If a vigorous yearling or two-year-old cock bird is used, twenty hens are not too many. On an egg farm where 2,000 to 5,000 layers are kept, an acre or two should be devoted to the breeders. This same plan of houses will serve all purposes, but instead of putting fifty hens to each pen, fifteen to twenty and one male bird should be the limit. Thirty to forty hens with two male birds will not do so well, unless the yard is planted with rows of corn or contains shrubbery. Where the fowls used for breeding purposes have considerable

10—A GOOD TYPE OF COLONY LAYING HOUSE

he could not keep them dry until he put in the board floors. During the winter time dampness is sure to come up out of the ground into the poultry house, being drawn out by the warmer atmosphere of the house. One of two things should be done: either fill in four to six inches of cinders or dry gravel and put three or four inches of fresh earth on top of this material, or use board floors.

Leghorns, or any other variety kept for winter eggs, should should not be allowed the freedom of the yards during stormy days or when a bitter cold wind is blowing; in fact we should not let them out even on sunshiny days when the temperature registers below ten degrees above zero. Leghorns' combs will freeze in bright sunshine when the thermometer is six above zero. Where the scratching and shed plan is used canvas curtains are placed in front of the scratching sheds, inside of the wire netting, and on cold stormy days the fowls are let out into these sheds. Here they are sheltered from the wind and much of the cold. Protected in this way, they will lay when fowls kept in the ordinary way will not lay an egg.

Should it be thought necessary to provide greater warmth, this can be done by building a cover over the roost poles a few inches above the heads of the fowls and dropping a curtain down in front, reaching below the roost poles, this curtain to be thrown back out of the way during the day-time and again

range, with shrubbery or other obstructions to an unlimited view located thereon, they may be allowed to run in large flocks, as they do on an ordinary farm, accompanied by a half dozen or more male birds; but in confinement in small yards, where all the birds are in view of one another, this plan will not give the best results. Mr. Wyckoff does not get nearly as many eggs during December, January and February as he does during March, April and May, but he gets a number, even during the winter months. One year his six hundred hens averaged one hundred and ninety-four eggs per hen, a very good record for so large a number of layers.

The fences used by Mr. Wyckoff are built in the ordinary way with 2 by 4 stringers, to which are nailed 1 by 4 pickets sharpened at one end. These pickets are six feet high and the fence is the same height. Many will be surprised that Mr. Wyckoff's hens keep their places when separated by a simple low fence of this kind. As pullets they do not keep their places so well, but as hens they become wonted and cause little trouble. It is not often one of the hens leaves her yard, but if she does, no damage is done except that Mr. Wyckoff prefers to have an equal number in each yard and house. Pullets of all varieties of fowls cause far more trouble than hens in the matter of flying fences. They are much lighter in weight and perhaps more giddy. As hens they become decidedly

EGGS AND EGG FARMS

heavier and more sedate in their manners. If difficulty should be experienced in confining the fowls to their runs, a single wire may be strung about six inches above the top of the fence. It should be so small as to be nearly or quite invisible to the fowl from the ground. If she flies for the top

31—ENJOYING A WIDE RANGE.

of the fence, as they generally do, to alight upon it, the single wire will prove an obstacle to any further attempts to get out.

Use your fowls well. Do not excite them, and you will experience little difficulty in this connection. It is the flighty, excitable layers that fly the fence on the slightest provocation.

GREEN FOOD IMPORTANT

The noon meal consists of the green food to be fed for the day. During the winter months he feeds mangel wurzels and cabbage; during the summer time clover and kale. He raises these foods in sufficient quantities. He gets clover started as early as possible, and after it is three to five inches high, feeds it until mid-summer weather burns it up; then he begins on kale, which renews itself and lasts him until freezing weather kills the plants. Kale looks like a cross between beet tops and pie plant. We do not know how better to describe it. The leaves look like pale green beet leaves, but are much larger. As the outer rows of leaves are picked off new ones come out of the heart of the plant, and keep on doing so until winter killed. It is certainly a great boon to poultrymen. Mr. Wyckoff had, when we were there, a patch about 25 feet wide by 150 feet long, which gave him all he needed for his 1200 fowls and chicks.

The mangels used by Mr. Wyckoff are the large variety commonly fed to stock. They are used by him during the winter. He runs them through a cutter, slicing them up as fine as he can and feeding them in troughs. The cabbage is fed in much the same way. Even in the matter of green food, Mr. Wyckoff aims to give his fowls only as much as they will eat up readily. He wants them to eat their fill, but has nothing for them to waste.

In connection with the noon feed of green food some whole grain (not much of it) is scattered in the litter in the houses to "work" the fowls more or less. The green food, however, is their main diet at noon, and Mr. Wyckoff lays great stress on its value as an egg-producing food. Said he: "It seems to me that I would almost rather stop feeding grain than green food. That is, of course, an extreme statement, as green food is mainly an appetizer and bowel corrective, but I could

not do business without a daily ration the year around of green food."

The evening meal with Mr. Wyckoff's fowls consists of mixed whole grain as follows: Two bushels of wheat, two bushels of oats, two bushels of buckwheat and one of corn. This is the proportion for summer feeding; in the winter time he increases the corn to two bushels, thus using equal parts of the four grains. As a variety, he feeds barley with the above, if it is low-priced. All grain is fed in litter. He prefers that they shall pick up at night all he feeds them, so they will meet him in the morning with sharp appetites.

Salt is given now and then in the soft food, though buttermilk is used each week along with the skimmed milk and sour milk to moisten the soft food, and this buttermilk contains salt. During the winter time boiling water is used to moisten the soft food whenever milk is short. This swells the food before it enters the crop, but makes it only barely warm, not hot. Hot, steaming food in winter time is a mistake, as it opens the pores of the skin unnaturally and subjects the fowls to sudden colds, with roup in prospect. Said Mr. Wyckoff: "I want the soft food as dry as I can get it and still be able to say it is moistened." Sloppy food of any kind loosens the bowels and brings on debility.

PRIVATE TRADE IN EGGS

WHY THIS TRADE IS DESIRABLE—WHAT IT DEMANDS—EGGS MUST BE STRICTLY FRESH AND ATTRACTIVE—HOW TO BUILD UP AND RETAIN TRADE AT GOOD PRICES

JOHN H. ROBINSON

THERE are two reasons why a market poultryman who wishes to sell his goods direct to consumers and thus retain for himself all possible profits, must produce extra choice articles. First, because it costs as much to produce poor eggs and poultry as it does to produce stock of superior quality, and costs more—takes more time and trouble—to sell the poor stock; and, second, because unless he does

32—COLONY HOUSES ON FREE RANGE

produce extra choice goods he can not get and hold the trade which buys good poultry and eggs freely and is willing to pay the price for them

In eggs the demand of the best trade is for a strictly fresh article of good quality, medium to large in size, with the color of the shell a matter of indifference in all but a very few markets. If your trade prefers eggs with shells of a particular color, dark brown or pure white, it is good business policy to cater to this demand. Never make the mistake of trying to force or persuade the trade to take something which it does not want.

The trade wants simply fresh eggs. In winter an egg less than a week old will readily pass as strictly fresh. In summer an egg more than three or four days old begins, usually, to be a little stale. As the sale of eggs is the mainstay of a private trade in poultry products, deliveries must be so timed as to keep customers always supplied with fresh eggs. There will then be two regular deliveries each week in summer and a weekly delivery in winter, and it will be found that this system is also well adapted to the needs of the dressed poultry branch of the business. But eggs which have no merit but their freshness will not suit this trade. To please it an egg must be of good quality. Fresh eggs can be procured from any hens that will lay at all, but nice, rich eggs come only from hens in good condition, and well-fed. To keep a large

to a person whose custom is seemed worth while to make a special effort to secure, always, of course, giving them to understand that the special price could not be obtained on future purchases. This little scheme was a good trade-getter.

As the poultryman's trade, like the order-wagon trade, of the grocer and butcher, is solicited at the kitchen door, it is well for him to get his dealings at each house on the right basis at the start and keep them there. When he first solicits custom at a house his errand should be to the mistress of the house, and if any fault is found with the goods he should ask to see her for explanation or adjustment of the matter. Many housekeepers give the purchase of supplies their personal attention, but often the girl in the kitchen is the active agent in the buying of table supplies, and, if she is not too honest, or if she wishes to favor the grocer—who also has eggs to sell, and who occasionally makes her a small present—or if she has a tender interest in the driver of the meat wagon, it is the

13—EFFECT OF SPECIAL FEEDING TO HASTEN THE MOLT

stock of hens, producing a steady supply of eggs quite uniformly up to the standard of the trade we are considering, is no light test of a poultryman's ability.

WORKING UP A BUSINESS

It is not hard for a poultryman who is producing what the best trade wants, to work up a good route of profitable, prompt paying customers, who will stay with him through all seasons, year in and year out. It is not hard, but it takes tact, patience, time, and, above all, some diligent soliciting. If one has read and remembered the old saying, "All things come to him who waits," now is a good time to forget it. Those who wait for this trade do not get it. It is the special perquisite of those who go out and hustle for it. As has been intimated, the way to get it is by personal solicitation. There may some times be an exceptional case, where it will pay to advertise for this trade. As a rule, advertising would not pay, because the volume of business is small, because the expense would be too great in proportion to the probable returns from advertising, and because it is desirable that the route should be kept as compact as possible. Our experience in advertising in local papers for this trade was that our advertisement appealed most to persons we did not want as customers, because we could not deliver to them without extending our routes more than the amount of their purchases would justify us in doing.

Our best drawing card when soliciting custom was this: "Take a dozen, or two dozen, or as many as you want of our eggs. Try them. If you don't find them better than what you are using you need not pay a cent for them. If you find a bad egg in a lot bought from us, we will give you a dozen fresh eggs for it." In selling poultry we would sometimes make a special price (the regular market price) for a trial purchase,

easiest thing in the world for her to give your trade a black eye. Right here is where many find the family trade in poultry products disagreeable. It is not always pleasant to have to do business under the conditions described, and one would, perhaps, rather let the matter drop and lose his customer than go to the trouble of putting things straight. But he ought not to let such a matter pass without being made right. He can not afford to let a customer quit dissatisfied with his goods and displeased with him when there is no ground for dissatisfaction, and the displeasure should fall elsewhere. After good goods, pleased customers are one's best advertisement, and after poor goods, the lost customer who thinks that he has a grievance can do a trade the greatest injury.

One important consideration in soliciting trade must always be your ability to supply customers regularly and fully. The agreement to do this will often secure trade when other considerations fail. Such an inducement, however, ought never to be offered unless one is sure of being able to carry it out. It is not a wise thing to do to take on all the regular customers you can get at the season of the year when your plant is producing best. If you do you soon find it necessary to drop some of them, and in that case it may be difficult to get their custom again when you would possibly be in a position to handle it permanently.

Finally, it is only in rare cases that a poultryman can build up a good private trade in eggs and poultry if he attempts to handle the products of others. If he knows his business as he ought to, and produces genuinely choice stuff, he will find that not one person in ten of those from whom he could buy has goods equal to his own. He will find also that, like himself, those who have good stuff to sell want all the profit there is in it.

REMUNERATIVE PRICES FOR EGGS

REPORT OF MANAGER OF POULTRY DEPARTMENT, OTTAWA, CANADA—RAPIDLY GROWING INTEREST IN POUUTRY CULTURE—EGGS SHOULD BE UNIFORM IN SIZE AND IN FLAVOR—HOW EGGS SHOULD BE SHIPPED—RATIONS IN EGG PRODUCTION

A. G. GILBERT

THE recent operations here have been successful beyond the average. There has been a marked and gratifying increase in the number of farmers who are giving their poultry proper care and management, so as to make them revenue producers. In a letter lately written by Mr. David Moir, a farmer near Almonte, Ontario, and a director of the North Lanark Agricultural Association, he says: "There has been more money spent for lumber and tar paper, wherewith to build poultry houses, in a few months past, than in a number of years."

Among the subjects treated in this report are the different markets for eggs; the cause of so many bad eggs being placed on the market; how to prevent bad eggs from being placed on the market; the result of different rations in egg production, and other matters which it is hoped will be interesting and instructive to the farmers and the poultrymen of the country.

The laying stock during the molt was carefully looked after. No attempt was made to stimulate egg production during that period. The hens, however, were fed a generous diet, in order to induce the growth of new feathers, and they had the run of a grass and clover field in the rear of the main poultry building. As soon as they were completely over their molt they received a liberal allowance of cut bone, and winter laying had fairly commenced by the end of November. As in previous years green cut bone was found a valuable incentive to egg production, and also beneficial, in smaller quantities, during the molting period.

34—NEST BOX FOR SITTING HEN

THE BRITISH MARKET PRICES UNLIMITED

That our farmers are beginning to realize the value of their poultry as money makers, is evidenced by the increasing demand for information as to the proper care and management of their birds, as well as by the increasing number of new-laid eggs placed on the market in recent winters. It may be said that if a greater number of eggs are being placed on the winter market, there will soon be enough to supply that market. Granted that there has been a greater supply of new-laid eggs in recent winters, there is also the fact that prices were never higher in Ottawa and Montreal than they were last winter, which goes to show that if there had been a greater production there has also been a correspondingly increased demand.

Observation and experience of the market in recent years led to the conclusion that the winter market is not the only paying one, but that there is a great and growing demand in the summer months for new-laid eggs of unimpaired flavor.

As for the English market, it is practically unlimited. A bulletin issued from the finance department states in effect that an unlimited, steady and profitable trade can be done with England in Canadian poultry and eggs.

The following information concerning the requirements of the British egg market is from the report of the Dominion Commissioner of Agriculture:

"The grade of egg which is in good demand in Great Britain is one weighing 15 pounds per great hundred, that is, 15 pounds per ten dozen, which is equal to two ounces per egg, or 1½ pounds per dozen. A small quantity imported into Great Britain from France go as high as 17 pounds per great hundred. For every half pound which eggs weigh less than 15 pounds per great hundred the value is lessened by about one cent per dozen.

"Eggs should be graded as to size. A higher value will be obtained for a given quantity of eggs graded into three sizes—large, medium and small—than if they are sent with sizes mixed promiscuously. Eggs of a brown shade of color are preferred.

"The preferred size of egg case for export is a wooden case holding thirty dozen eggs, paper filled—that is, having pasteboard frames with a separate space for each egg. These cases, holding thirty dozen each, measure about 28 inches long by 12¾ inches wide and 13 inches high outside.

"For the safe carriage of the eggs, it is important that they should not be stored in a warehouse, on the cars, or on board the steamship, in proximity to any cargo from which they would acquire a flavor. The carrying of eggs with a cargo of apples has been known to impart to them a flavor which impaired their value.

"They should be carried on the cars and on the steamship at a temperature of from 38 to 42 degrees Fahrenheit. When cases containing eggs are removed from the cold storage chambers, they should not be opened at once in an atmosphere where the temperature is warm. They should be left for two days unopened, so that the eggs may become gradually warmed to the temperature of the room where they have been deposited. Otherwise a condensation of moisture from the atmosphere will appear on the shell, and give them the appearance of sweating. This so-called "sweating" is not an exudation through the shell of the egg and can be entirely prevented in the manner indicated."

Eggs that are placed in cold storage from April till July are shipped to Great Britain for the September and October trades. Eggs that go into cold storage in the fall are exported during the winter months. Cold storage eggs are sold in Great Britain as "Canadian fresh eggs," and the prices have ranged from 7s. to 7s. 6d. per long hundred (120 eggs) during

EGGS FOR PROFIT

September and October, and from 8s. to 9s. per long hundred during November and December.

Pickled eggs should be exported to Great Britain so as to reach there during November and December. The eggs that were sold during November and December last year realized 7s. to 7s. 6d. per long hundred.

"There is undoubtedly a growing inclination among consumers to give preference to Canadian eggs for winter trade, and the shipment to the United Kingdom may be very largely increased without injuring consumption, provided always in the first place that the quality is maintained up to a high standard; and, secondly (a most important one for Canadian shippers), that the price is not prohibitive."

REMUNERATIVE PRICES IN CANADA

In proof of the high prices of winter, it may be stated that the writer attended an agricultural meeting in Montreal during one January, when he was informed by several farmers present that they had sold new-laid eggs the week previous at 60 cents per dozen to choice customers. It is but right to say that at the same time new-laid eggs were selling at 35 cents per dozen at retail in Toronto and 25 cents per dozen in London, Ontario. In Manitoba and the northwest prices ranged from 35 to 50 cents per dozen, according to locality. Mr. Sutherland, assistant secretary of the Montreal Poultry Association, wrote later on that he had sold his new-laid eggs during that winter at the first named price. Eggs at 60 cents per dozen meant that they were a luxury which only the rich could indulge in. If eggs were put on the Montreal market during the winter in such numbers that lower prices would follow, it is only reasonable to suppose that more people would purchase them. There is no reason why the great masses should not be supplied with new-laid eggs in winter rather than the ill-flavored, artificially preserved article, at a price within the reach of all, and there yet remain a paying margin of profit to the farmer. In order to find out what are remunerative figures, the summer market prices, at about their lowest points, viz., 12 to 15 cents per dozen, are taken. The following calculation is made, based on the experience of several practical breeders:

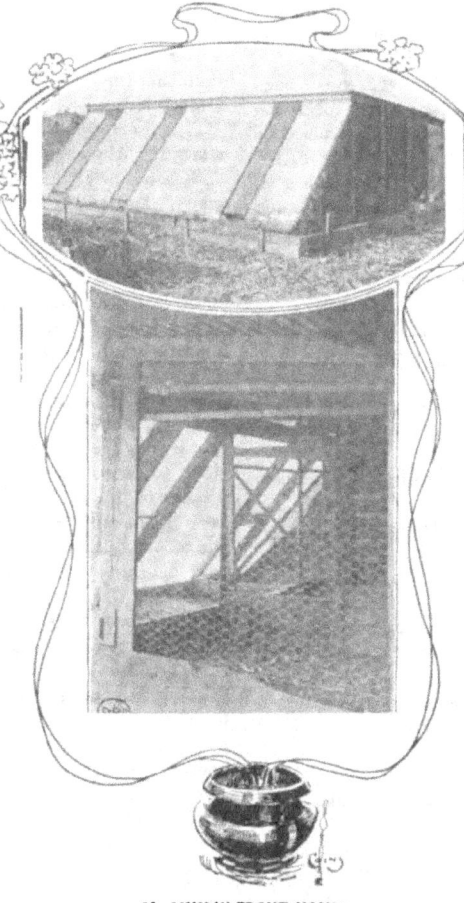

35—MUSLIN-FRONT HOUSE

100 eggs from hen for one year, at 1 cent each	$1.00
10 chickens hatched by her, at 10 cents each	1.00
Body of hen to sell or eat	.25
	$2.25
Deduct cost of hen for year	1.25
	$1.00

We have, according to the foregoing, a margin of $1 per hen profit per annum, taking eggs at 12 cents per dozen. No figure is placed upon the manure, which is valuable when made into a compost. It may be said that the cost of producing the egg is greater in winter. But this statement may be met by the other that the cost of production is little in summer, for at that period the farmer's hens, in most cases, are allowed to forage for their living. So that the cost of $1.25 per hen per annum is very fair—if anything, it is on the high side. It will be seen that eggs, at the summer price of 12 cents per dozen, afford a paying margin. Surely, then, with the modern and cheaper rations, prices during the winter season should be much lower, and yet afford a fair profit.

But the summer price of 12 cents per dozen is a misleading one, for in reality it should be placed at twice the figure. Twenty-four cents per dozen for eggs in mid-summer? Yes, and in this way: It is a well-known fact that during the midsummer months it is hardly possible to buy from farmer or storekeeper a dozen or two eggs that will all be found good; in the majority of cases half of the eggs will be likely unfit for eating purposes, making the six actually worth 12 cents, or 24 cents per dozen, and probably the flavor of the remaining six will not be such as new-laid eggs ought to have.

There is not the slightest doubt that the great majority of purchasers would rather pay 24 cents per dozen, in the first place, for a reliable article than half the amount for inferior goods. There is no intention to say that our farmers bring into the market, or sell to the dealers, or that the latter dispose of bad or ill-flavored eggs, knowing them to be such. On the contrary the farmers, as a rule, unfortunately give as little attention to the age or condition of the eggs they are taking to market as they give to the fowls which laid them. The question may be asked, How can we tell what the inside of an egg is like? How can we distinguish the bad eggs from the good ones?

The answer to the above queries is that while the farmer is not supposed to be in the van of poultry lore as to the means of discovering partially hatched or ill-flavored eggs from the new-laid ones, yet there are simple precautions which may be taken in order to secure the new article, and which he is in duty bound, in the interest of his customers, to take. By observing the following, eggs of fine flavor may be sold during the entire summer season:

PRACTICAL ADVICE TO FARMERS

1. Keep no male bird with the laying stock.
2. Collect the eggs once or twice every day.
3. Take no eggs to market gathered from under barns, nests in the fields, or from stolen nests.
4. Prevent, if possible, the laying hens eating decayed vegetable or animal substances.
5. Keep the eggs, after gathering them, in a cool, sweet atmosphere. If in a cellar, let it be dry.
6. Keep the nests the layers use clean, comfortable and free from vermin.
7. Have a sufficient number of nests for the layers.

Offer every inducement to the hens to lay in these nests and not shun them.

8. Allow no brooding hen to sit on the new-laid eggs, be it for ever so short a period.

9. Take the eggs to market clean and inviting in appearance.

10. Make it a rule to take no eggs to market that you are not sure are fresh, or that you are doubtful about the flavor being good.

The question is frequently asked, and much speculation indulged in, as to where all the bad eggs come from, particularly in summer time? And that leads to the question, What is a bad egg?

In the past eight years large numbers of eggs have been handled in our poultry houses. Many eggs have been put under hens, or in incubators, and close observation has been made of these eggs during incubation, and afterwards of the eggs which failed to produce chickens. The eggs in course of incubation were also tested at the end of six or seven days and note taken of the varied appearances presented. No small amount of experience was gained, and it leads to the classification of the different sorts of eggs met with, and the cause therefor, as follows:

1. The fertile egg, in which the germ is in a well-advanced stage, with the promise of making a strong, vigorous chicken.

2. The addled egg, or one in which the germ has started but for some cause its progress has been arrested, when decay sets in and you have a very ill-flavored article.

3. The clear or infertile, which contains no germ and presents the appearance of a new-laid one.

4. The egg containing a broken or ruptured yolk, and which presents a similar appearance to No. 2.

The state of Nos. 1 and 2 can only result from fertilization.

No. 2 is the egg most frequently met with, and is probably the result of taking eggs from nests under barns, or stolen nests, or nests on which the hen has been sitting some days.

No. 3, the clear or infertile egg, can be used for cooking purposes with every confidence after examination by tester on the seventh day. The infertile eggs are frequently removed after the fertilized eggs have hatched (on the twenty-first day); then they are boiled hard and fed to the chicks.

Having secured the non-fertilized, new-laid eggs, care should be taken to preserve the flavor intact. The shells of the eggs are porous, and contaminating surroundings will doubtless affect the egg. The unfertilized egg may be kept in a cellar, with pure atmosphere, for many weeks and yet retain its flavor. In course of time it may shrink and partially dry up from evaporation, but there is no germ to start on its mission of bringing about change as soon as the conditions are favorable, or partly so. The unfertilized egg will keep longer than the other, and, it is said, an egg from a hen fed on corn will keep its flavor better. All eggs should be kept in a sweet atmosphere.

It must be borne in mind that it is the flavor of the egg that is all important to keep intact. And on this point a farmer in the neighborhood of New York City who sends thousands of eggs per week to that city, writes to the Rural New Yorker, that "if a brooding hen is allowed to sit on a new-laid, fertilized egg for twelve hours the flavor of that egg is ruined." The same authority, who uses a large number of incubators, says that he tests his incubator eggs on the fifth day, and all the clear or infertile eggs he removes, marks them as such, and ships them to New York City, where they are sold for cooking or baking purposes.

In our poultry department eggs have been tested on the sixth and seventh days and the infertile eggs have frequently been boiled hard wherewith to feed the chicks. On some occasions, at the end of the hatching period of twenty-one days, the clear or infertile eggs have been removed from the nest and boiled hard to mix with chicken food. All poultrymen know that it is impossible to boil a rotten egg hard.

It must not be inferred from the foregoing that unfertilized eggs should be kept a long time before being taken to market. Eggs should be sold as soon after being laid as possible. There are cases where the farmer is some distance from the purchaser or can not come to market as frequently as one nearer to the city. In such a case, the eggs for sale may have to be kept for some time, and it is all the more important then that they should be unfertilized and kept in a cool, sweet atmosphere.

In the opinion of the writer it is only a matter of time and education when eggs for sale in summer will have to be guaranteed as unfertilized by the seller before a purchase will be made. Indeed the subject is already receiving practical attention. A prospective question likely to be asked, in connection with its discussion, may as well be answered, viz. If we are to allow no male bird with the laying stock how are we to breed our chickens? Easy enough, by picking out in early spring time, or better still, if circumstances will permit, by keeping apart all winter and not stimulating them to lay—nine or eleven of your best layers and best shaped birds. Mate them with an unrelated, healthy, well-shaped two-year-old cock if the birds are pullets or yearling hens, and a cockerel if they are two years old. When eggs enough have been saved to hatch out what chickens you wish, close up, kill or dispose of the male bird, and after keeping the hens he has been mated with inclosed for a week longer, let them run with the other laying hens, with which there is, of course, no male. And having saved eggs for hatching from birds selected for good qualities, superior progeny is likely to follow. The chickens from eggs saved from such mating will certainly be better, in every way, than those bred in the usual hap-hazard manner. As to keeping the male bird with the laying stock, the following is again quoted from Experimental Farm Poultry Department report: "The cock bird is a nuisance in the pen of layers. He not only monopolizes most of the food, but teaches the hen to break eggs and so learn to eat them. Besides, the stimulating diet is too fattening for him and will ruin him as a breeder."

CONCLUSIONS PROVE IT A PROFITABLE BUSINESS

In noting, in the foregoing, the features of the different markets, the demand and supply peculiar to them and the requirements of the various seasons, the following conclusions may be arrived at, viz:

1. That our home winter market offers the inducement of high prices for new laid eggs.

2. That notwithstanding greater production in this district, prices were never better than now.

3. That there is no reason why new-laid eggs should not be produced, in winter, in such quantity as to take the place (in a very great measure) of packed, or preserved eggs.

4. That with the modern and cheaper rations in vogue, winter prices could be much lower than they are and yet afford a profitable margin.

5. That eggs in the summer months that can be relied on as being new-laid and of good flavor, will bring better prices than the ordinary article.

6. That so many summer eggs are bad, or ill-flavored, because (a) they are not unfertilized; (b) not collected immediately after being laid; (c) not brought to market soon after being laid.

BREEDING FOR EGGS

PEDIGREE BREEDING FOR EGG PRODUCTION

THE GREATEST EGG PRODUCER IS THE FOWL THAT HAS BEEN BRED FOR THE SOLE PURPOSE OF PRODUCING EGGS—SUGGESTIONS FOR HOUSING, FEEDING AND HATCHING TO PRODUCE THE LAYER—EARLY HATCHING OF CHICKS IS ADVISED

ROBERT H. ESSEX

THE greatest egg producer is the bird that has been bred for the sole purpose of producing eggs. This bird will not necessarily be a Leghorn or a Minorca, although these breeds deservedly have the reputation of being the greatest egg producers living—that is, as a class. Without doubt there is a greater proportion of eggs laid by these two breeds than by any other two breeds that can be named; yet there are many individual birds of other breeds that may equal or even surpass them. If such should be the case, it will be found that these individual birds have been bred with one object in view, namely: egg production. Just as the fancier raises birds for exhibition, so may the farmer breed birds for laying purposes. Undoubtedly the proper course to pursue would be to choose your prospective layers from a class already noted for their laying proclivities, but do not imagine that you have the best layers on earth simply because the breed selected has that reputation. Many Minorcas and many Leghorns have proved unsatisfactory layers, while many a Brahma and many a Plymouth Rock has abundantly helped to fill the egg basket.

As I have said, if you are commencing, select your birds from the classes bearing reputations as layers, but do not be discouraged because it is not convenient to do this. You may commence now with the stock in hand and note the best layers among your birds. Begin line breeding with as great regard to mating as you would if breeding for show purposes. Mark the pullet that is the first to lay; mark the most persistent layer; mark the hen that molts quickly and gets down to business before the hard winter sets in; mark also the best winter layers, and when you have done marking, the spring will be here and you may commence mating. Better to breed from two or three well known layers than to take chances and make up a pen containing a dozen indifferent ones. As the cock does not lay, you cannot judge whether he is likely to produce good layers, unless you know how he is bred, but you can choose the largest and most vigorous bird of the flock to mate with your selected females. After that it is easy. Never allow anybody to induce you to change the blood of your flock by the introduction of a male bird of another strain, unless you are satisfied he comes from a strain which equals your own as layers. Remember the sire controls one-half the blood of the produce, and if you desire to introduce new blood or new stamina into your flock do so by means of the best laying female you can procure. Even then I would not use her sons as sires, but would dispose of them and mate her daughters back to the old male bird; the produce of this mating would have in their veins three-quarters of the blood of your own strain, with sufficient new blood to maintain the vigor of the flock. Do not overlook the necessity for observation each year, so as to intelligently mate your birds the next season, continually choosing the best layers and limiting your breeding pen to these. The result will be that no matter what breed you start with you will eventually own layers far ahead of any that have been indiscriminately bred. The same advice applies to production of large eggs. I have had Minorcas which have laid large eggs, and Minorcas which have laid small eggs; Brahmas, layers of large eggs, and Brahmas, layers of small. During recent years in breeding Buff Plymouth Rocks I have found that some hens lay small eggs, others large; and as I have carried out the system of pedigree breeding I have noticed the fact that layers of large eggs transmit this attribute to their progeny, and layers of small eggs have produced birds which have also laid small egg. It rests altogether with the particular strain of birds, and not with the breed, as to which will give the best return, either in size or number of eggs.

There is a material difference between 150 eggs a year, which is a fair average, and 289, which is I believe the record of a pen of fowls which have been entered for competition in an egg-producing contest. It shows what can be done by pedigree breeding and judicious feeding; and constitutes the difference between profit and loss.

If you keep many varieties you cannot give the necessary time to each one. Since I limited myself to breeding Buff Plymouth Rocks I have won more prizes and obtained more satisfaction than I did on all the others combined.

HOUSING, FEEDING, HATCHING

There has been so much information given as to raising, housing and feeding, that anybody who reads should have no difficulty in these respects, if the directions are faithfully fol-

EGGS AND EGG FARMS

lowed. Each breeder may have different methods, but analyzed they will be found to agree in the main. One feeds cut green bone every day, another every second day, but the amounts fed also differ, and the result is much the same. One feeds soft food for breakfast, another for dinner. Even this is regulated by the habits of the poultryman. The man who feeds early in the morning may with good results feed grain as a breakfast while the one who feeds late will do better by giving the soft food first. The hens become habituated to certain methods, and will do fairly well under any, so long as they are not too radical. Still, the man who gets up early and feeds his fowls regularly will get the best returns, and he deserves them.

Give little soft food, a small but regular supply of meat, or ground green bone, and a variety of grain, not forgetting the green food in winter, and the principal requirements for egg production have been performed. The next important requisite is work. Feed the grain in litter, cover it well and make the hens work to find it. Do not be governed by false kindness, and throw down the food in heaps, but cover every grain. Be careful as to exciting the birds. Strange dogs, cats or even your next door neighbor going among the hens when in confinement will affect the layers detrimentally. A change of pens, removing a hen from one pen to another will cause a cessation of laying for a time. Change the position of your nests, and it has the same effect. Introduce a strange male bird, and you will notice the reduced number of eggs. Any change, every change should be guarded against.

Give plenty of room and plenty of sunshine to the workers, and never reduce the scratching space to less than six or eight square feet per hen. Even this amount is small, and when confined to such a space it is necessary to limit the number of fowls in a pen to ten or a dozen. The most important requirement has not been mentioned, that is the water. Watch the hen come off the nest after laying and see her make for the water, and you will understand the necessity for pure water and lots of it.

In the winter, if your house is dry, the fowls will keep themselves warm during the day if you feed little and often, and make them work. At night care must be exercised to see that they have a warm corner for a roosting place.

Hatch your chicks as early as possible, but certainly not later than May, and if properly cared for you will have winter layers, and receive all the way from 25 to 50 cents a dozen for your eggs. If you allow the hen to use her own sweet will she will probably incubate in June, July and August, and you will have lots of worry, lots of squabs, and any amount of expense feeding during winter chicks that bring you no return.

EGG YIELDING CAPACITY OF HENS

USE OF TRAP NEST SHOWS GREAT VARIATION IN EGG PRODUCING CAPACITY OF HENS—SIX YEARS' INVESTIGATIONS IN RELATION TO BREEDING FOR GREATER EGG YIELD—PRACTICAL WORK BY MAINE AGRICULTURAL EXPERIMENT STATION

PROF. G. M. GOWELL

IN breeding poultry such wonderful changes have been made in form and feather that it seems to have been demonstrated that the laws of inheritance and transmission are as true with poultry as with cattle, sheep and horses. Many attempts have been made to improve egg production by breeding. This work has, for the most part, been by flocks rather than by individuals, much the same as if we would attempt to improve the milk production of the herd by basing the breeding upon the milk or butter production of the herd without reference to the milk of the individual animal.

Our work is based upon the individual records obtained by the use of trap nests devised by us. The houses are specially constructed for the purpose, and while numerous problems have arisen in connection with the work, and other side questions have been studied, nothing has been allowed to interfere with the original proposition of breeding for increased egg production. At the time we began this work we were carrying three breeds—Barred Plymouth Rocks, White Wyandottes and Light Brahmas. With the particular strains that we had of these breeds, the Barred Plymouth Rock seemed the most promising and the work here reported is with this breed. As the New England market demands large, dark-brown eggs, only birds laying such eggs have been used in the breeding.

FIFTY-TWO TRAP NESTS WERE CONSTRUCTED

In recognition of the necessity for improvement of the egg producing capacities of hens, the work was taken up, and November 1, 1898, two hundred and sixty April and May hatched pullets were put into breeding pens and records kept of their individual productions for a year. This was done with much certainty by use of the trap nest boxes described as follows in the station report for 1898:

"We constructed a nest that proved so satisfactory that we constructed fifty-two of them in the breeding house. * * * The boxes are placed four in a bank and slide in and out like drawers, and can be carried away for cleaning if necessary. If desired they could be put on the floor or shelf by simply adding a cover to each box. * * * To remove a hen the nest is pulled part way out, and as it has no cover she is readily lifted up and the number of her leg-band noted on the record sheet that hangs at hand. * * *

"The nest box is very simple and inexpensive, easy to attend and certain in its action. It is a box-like structure without front end or cover. (See illustration on page 46.) It is twenty-eight inches long, thirteen inches wide and thirteen inches deep, inside measurements. A division board with a circular opening seven and one-half inches in diameter is placed across the box twelve inches from the back end and fifteen inches from the front end. The back section is the nest proper. Instead of a close door at the entrance, a light frame of inch by inch-and-a-half stuff is covered with wire netting of one-inch mesh. The door is ten and one-half inches wide and ten inches high, and does not fill the entire entrance, a space of two and one-half inches being left at the bottom and one and one-half inches at the top, with a good margin at each side to avoid friction. If it is filled the entire space it would be clumsy in action. It is hinged at the top and opens up into the box. The hinges are placed on the front of the door rather than at the center or back, the better to secure complete closing action.

"The trip consists of one piece of stiff wire about three-

BREEDING FOR EGGS

sixteenths of an inch in diameter and eighteen and one-half inches long, bent as shown in the drawing. (See illustration on page 46.) A piece of board six inches wide and just long enough to reach across the box inside is nailed flatwise in front of the partition and one inch below the top of the box, a space of one quarter of an inch being left between the edge of the board and the partition. The purpose of this board is only to support the trip wire in place. The six-inch section of the trip wire is placed across the board and the long part of the wire slipped through the quarter-inch slot and passed down close to and in front of the center of the seven and one-half inch circular opening. Small wire staples are driven nearly down over the six-inch section of the trip wire into the board, so as to hold it in place and yet let it roll sideways easily. When the door is set the half-inch section of the wire marked A comes under a hard wood peg, or a tack with a large round head, which is driven into the lower edge of the door frame.

The hen passes in through the circular opening and in doing so presses the wire to one side and the trip slips from its connection with the door. The door promptly swings down and fastens itself in place by its lower edge, striking the light end of a wooden latch or lever, pressing it down and slipping over it, the lever immediately coming back into place and locking the door. The latch is five inches long, one inch wide and a half-inch thick, and is fastened loosely one inch from its center to the side of the box, so that the outer end is just inside of the door when it is closed. The latch acts quickly enough to catch the door before it rebounds. * * Strips of old rubber belting were nailed at the outside entrances for the door to strike against.

"The double box, with nest in the rear end, is necessary, as when a bird has laid and desires to leave the nest she steps to the front and remains there until released. With one section only she would be very likely to crush her egg by standing upon it."

NO FIXED TYPE FOR LAYERS

A study of the monthly record sheets shows the great differences in the capacities of hens, and marked variations in the regularity of their work, some beginning early and continuing laying heavily and regularly month after month, while others varied much, laying well one month and poorly or not at all the next. Accounting for these vagaries was not practicable as the birds were bred alike, and selected for their uniformity.

All pens were of the same size and shape and contained

36—INDIVIDUAL RECORD NESTS IN POSITION

the same number of birds. Their feeding and treatment was alike throughout. Whenever changes were made in the feed in one pen they were made in the others. That they were in good health is shown by the fact that but two were ailing, and were taken out early; two were crop-bound, and one was injured by rough treatment by a cockerel.

Many of the lightest layers gave evidence of much vitality and in many instances there were no marked indications in form or type by which we were able to account for the small amount of work performed by them. Numbers 234, 70 and 236 yielded respectively 36, 37 and 38 eggs in the year. They were of the egg type and gave no evidence of weakness or masculinity. Numbers 101, 296, 36, 47 and 14 with their yields 204, 206, 201, 200 and 208 eggs during the year were typical birds with every indication of capacity, but they were equalled in the minds of good judges by other birds that yielded a much less number of eggs. The size and uniformity of the eggs yielded are of a good deal of importance. It was very noticeable in these investigations that the eggs from hens that laid the greatest number averaged smaller in size than from those that did not produce so many. That this was not always the case was shown by the eggs from numbers 101 and 286, which were of good size and dark brown, while those from number 36 were small and lacking in color.

When the eggs from the hens that had been laying long and freely were placed in the incubator, many of them were found low in fertility, or entirely sterile, notwithstanding the hens had mated freely with vigorous cockerels. The percentage of infertility was much greater than in eggs from hens that had been laying moderately. The question arises whether a large percentage of the chickens raised each year are not the produce of the tardy and moderate layers that are comparatively fresh, rather than of the more valuable and persistent layers that have been hard at work all winter? If this is so, breeding from eggs as they are ordinarily collected without a knowledge of the hens that produced them, can but tend to furnish a large proportion of chickens from the poorest hens in the flocks.

The cockerels as well as the pullets raised in this way furnish the breeding stock for the next year and in this manner the reproduction of the poorer rather than the better birds is fostered.

For twenty-one years I have been at work with the same family of Barred Plymouth Rocks, and by selecting typical eggs for incubating have succeeded in very much improving the shape, size and color of the eggs yielded. That was an easy matter, for I simply bred producers of quality in order to

EGGS AND EGG FARMS

secure quality, and I secured it. I endeavored to increase the egg yields by selecting birds of what I thought was the "egg type," and breeding them together, I had heard a great deal about the "egg type" and had gotten to think it was a hard and fast fact. After using the trap nests for a few years, however, and finding in the same pens, where the hens were all from the same hatch, and fed and treated alike throughout their laying year, some birds that yielded from 220 to 251 eggs, and others that laid only from forty to sixty eggs during the same time, and not being keen enough of sight and touch, to discover differences of form and feature, sufficient to account for the great variations in yield, I began to lose faith in the "Beef and Dairy Form," as indicating the internal functions of hens, sufficiently to be longer accepted as guides in selecting stock from which to breed egg producers.

RESUME OF EXPERIMENTS

In the laying pens for this test were placed 140 April-and-May-hatched pullets, which commenced using the nest boxes described November 1, 1898.

In one year forward from that date the 140 birds laid an average of 120 eggs each. Twenty-four laid over 160 each, and twenty-two less than 100 each. Hen No. 36 laid 201 eggs; No. 101 laid 204, and No. 286 laid 206 eggs. As the eggs of No. 36 were small and light in color, she was rejected.

TWO-HUNDRED-EGG HENS IN SPECIAL PENS

As from the start we had two birds that laid over 200 eggs, large dark-brown ones, in their first laying year, we have in our special breeding used no females that have laid less than 200 eggs in the first laying year, and in the breeding for general stock since the first year, only females that have laid from 160 to 200 large, brown eggs in their first laying year.

At the commencement of the next breeding season (1900) hens Nos. 101 and 286 were mated with males that were unrelated to them, or to each other. The cockerels raised from the eggs of these two birds were the first males produced for use in this work. In the early spring of 1901 several sons of these hens raised the previous year were mated with the twenty-four two-year-old hens that laid 160 egg or over each during 1899, and twenty-five others that laid 160 or over during the 1900 test.

That season hen No. 393 who had laid over 208 eggs during 1900 was bred to a son of 286. Hen 326 had laid 211 during 1900 and she was bred to a son of 286 also.

No. 318 had laid 237 good brown eggs in 1900. After she had laid 200 egg, the next twelve she laid weighed 1 pound, 11¼ ounces. She was bred to a son of No. 101 that season.

The sons of Nos. 101 and 286 were in service only during the year 1901.

ONE PULLET LAID 251 EGGS

In 1902 one hundred pullets were tested for additional foundation stock. They yielded an average of 132 eggs each. Twelve birds laid over 200 eggs each, the highest number being 251 eggs laid by hen No. 617. In the same pens were six others that laid only from 23 to 70 eggs each. Thirty-seven laid over 160 each. No hens were used as breeders that had not laid over 160 eggs, and all, as in the previous years, were bred to males whose dams had yielded over 200 eggs.

Males were raised in 1902 for the male breeding pens of the next year from hens No. 630, with a record of 213 eggs, and No. 676, with a record of 209 eggs. The eggs from both of these hens were very large and dark brown. They were mated to sons of No. 318 before spoken of. Males for the pullet breeding pens of the next year were bred from other matings of hens that had produced 200 eggs, with males whose mothers had yielded over 200 eggs.

RESULTS OF THE YEAR 1902

That year (1902) we were crowded for room and could accommodate only 53 pullets for testing. They were the first pullets that we tested that were sired by males bred from 200 egg-producing hens and show the first results of the breeding

37—SINGLE NEST REMOVED

practised. They had been laying quite heavily out in their summer quarters during September and October, although they were not hatched until April and May.

The 53 birds laid 7952 eggs in the year forward from November 1, 1902, a little better than 150 eggs each. Could they have been in quarters where their eggs could have been traced to them a month earlier, when they were laying so well, they would have shown a better year's work, as the twelfth month of their testing was really the thirteenth month of their laying, and the record sheets show it to be nearly bare of eggs. As it was, however, seven of the 53 show records of from 201 to 240 eggs each in the year, and 23 of the 53 laid over 160 eggs each.

THE 1903 SEASON

During the breeding season of 1903 hens No. 1001, record 213 eggs, No. 1003, record 240 eggs, No. 1005, record 222 eggs, and No. 1140, record 211 eggs, were bred to male birds raised the year before whose dams had yielded over 220 eggs each, for the purpose of procuring males for the male breeding pens of 1904. No two of these females were bred to the same male or to brothers.

All pullets raised that year (1903) were, as in the preceding three years, out of hens that had laid over 160 eggs in a year and they had the advantage over their predecessors, in that their dams and maternal granddams were sired by males from over 200-egg-yielding mothers, as they themselves also were.

A DISASTROUS WINTER MOLT

That year (1903), 160 pullets were tested in trap nests. They laid 21,202 eggs, an average of 132 each. Forty-four laid over 160 eggs each; nine laid 200 or over, viz., 200, 205, 210, 217, 220, 220, 221, 222, 225. We have not to seek far for an excuse for the lower yield than that of the last preceding year. The pullets were hatched in April and May, and thinking to have them mostly in

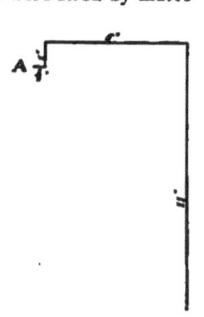

38—TRIP WIRE

readiness for laying early in the month of November, we fed them rather more beef scrap than usual during the growing season while they were out on the range, and before we were aware of their development they were laying in August. They were nearly all laying heavily during September, October and November. They were splendid birds

39—DEMAND WILL NEVER CEASE

but almost every one of them molted completely in December and we got very few eggs from them for more than two months. The most of the eggs secured from them were laid after the middle of January. Could the records have commenced September 1 and continued for a year, the showing would have been much better.

THE 1904 SEASON

The breeding season of 1904 opened with 170 yearling hens in our houses that had laid above 160 eggs each the year before; 80 pullets and hens whose mothers had laid over 200 eggs per year; and 28 hens that had themselves laid over 200 eggs per year. The birds were in twenty-four different pens, and they were bred to selected cockerels whose mothers had yielded above 200 large brown egg per year.

Among the pullets tested during the last preceding year (1903), were found the following: No. 263 yielded 220 eggs; No. 225, 220 eggs; No. 22a, 221 eggs; No. 224a, 222 eggs; No 205a, 225 eggs. These birds were mated during 1904 to different cockerels raised in 1903, and from 200-egg-producing mothers whose other sons were never used in our breeding operations. The matings of these five pairs of birds were to secure cockerels for our next year's breeding operations.

EGG-PRODUCTION INTERFERED WITH

At the usual time for the commencement of the yearly test of 1904, viz., October 31, 1903, we had 300 good pullets that were laying well out on the range. The construction of the building being erected for their quarters was interfered with by a question of labor, and they remained out in their small summer houses during a wet, cold fall and early winter until December 6, when they were moved in. This more than a month's delay and exposure cut into the year's work heavily and the average production of the 300 birds was reduced to 131 eggs each for 11 months. Nine birds yielded above 200 eggs each before the close of the following October.

All of the females carried in 1903-04 were tested hens that had laid from 160 to 251 eggs a year, and 150 pullets and hens whose mothers produced 200 eggs or over per year. The father or fathers, and the grandfathers of these two classes had mothers that laid 200 eggs, or over, per year.

THE PRESENT SEASON

This season (1904-05), 600 pullets out of hens that have laid over 160 eggs per year, and whose fathers, grandfathers and great grandfathers were out of hens that yielded above 200 eggs per year are being tested by the trap nest for additional breeding stock. All of the mothers of these pullets had fathers and grandfathers that had 200-egg-producing mothers.

The stock is strong and vigorous and but few chickens that hatch are lost. The hardihood of the stock is shown by the fact that many cockerels have been sold to farmers and poultrymen in and out of the state in the last two years, and this fall many of them have ordered again, with the frequent comment that their pullets are laying earlier in the season and giving better eggs than they had ever done before.

INBREEDING IS STRICTLY GUARDED AGAINST

The numbers of the breeding stock now secured makes practicable the avoidance of inbreeding, and this is strictly guarded against, as it is doubtful if the inbred hen has sufficient constitution to enable her to withstand the demands of heavy egg-yielding.

During only one season have birds as closely related as first cousins been bred together. Line breeding is followed, the matings being now only with distantly-related birds.

RESULTS OF THE SIX YEARS' SELECTION

These breeding investigations have been in progress six years. The first year was consumed in testing pullets to find foundation stock. The second year cockerels were raised from the heavy-laying hens for future breeding, and the third year the first lot of pullets were raised from the selected stock, so that we have only the last three years in which to note results, and these three years can only show first changes that have taken place. The stock that we first commenced with was pure and well-bred, as flocks go. The hens were averaging about 120 good brown eggs a year, and had been doing so for

40—COLONY HOUSES ON RUNNERS

several years. Two years ago they averaged 150 eggs and this year, and last, with the great setbacks caused as above indicated, which were no faults of the stock, the average was 131¼ eggs.

As the housing, treatment and food, have been as nearly

EGGS AND EGG FARMS

alike as we could make it during the last five years, there seems to be reason for assuming that the flock yields of 1902, 1903 and 1904 over those of previous years are the results of the breeding practised.

"ONE HEN IN SEVEN LAID ABOVE 200 EGGS"

Again, during 1902 about 1 hen in 7 laid above 200 eggs, while taking the three preceding years together but 1 hen in 28 laid 200, and the first year only 1 hen in 47 laid 200. It must be borne in mind that 1902 was the first year that we had pullets from the 200-egg stock to collect eggs from.

In the records only the eggs laid in the nests are accounted for. Had those found on the floor been reckoned in, the average per bird would have been slightly increased.

Sufficient time has not yet elapsed since the beginning of these breeding tests to establish claims of increased productiveness, but the outlook is certainly very encouraging.

The plans on which we are working are based on every day common sense. We are rejecting the drones, and breeding producers together to secure producers. It is known that the laws of inheritance and transmission are as true with birds as with cattle, sheep and horses, and, when we consider the wonderful changes that have been made in the form, feather and egg production of hens since their domestication commenced, there is ample reason for assuming that a higher average egg production than the present can be secured by breeding only from those birds which are themselves great producers.

The purpose of this work should not be misunderstood. We are not trying to breed stock that shall average to yield 200 eggs per year. If the average yield of the hens of the breed should be increased to the extent of a dozen eggs per bird, the value of this work would be many fold its cost.

HOW THE GREAT LAYERS ARE HOUSED

All of this breeding stock is housed in a long poultry house, which is one hundred and fifty by sixteen feet in size, eleven feet high in front and six feet high at the back. There are thirteen pens ten by sixteen feet and at the east end a feed cook room twenty by sixteen feet in size. In the south front are sixteen twelve-light windows of ten by twelve glass set in the upper half of the front so that the light is radiated over the center and rear of the pens, and below them are sixteen three-light windows of ten by twelve glass which light the space beneath the walk; all of these windows are double, the outside sash being hinged at the top so they may be swung out to permit an inflow of fresh air on mild, sunny days. Small ventilating spaces are cut high up in the front, between each two pens, and covered by low "hoods" to exclude snow and rain. These permit an outflow of the impure air of the interior at the same time that the slightly opened large windows give an inflow of the pure outside air, this adjustment of inflow and outflow securing perfect ventilation without drafts. A novel feature in this long house is the walk, extending along the fronts of the pens and about two feet above the pen floors. The view of the interior of this long house shows this walk and the gates opening into the pens; we also see that the large windows are placed high up in the front and their light extends clear across the sixteen feet of depth of the pens. The roost platforms and roosts, seven feet long, are at the back (north) ends of the pens, the platforms being three feet above the pen floors. The other three feet of this space is occupied by the double tier of trap nest boxes, above them being set a

41—HOUSES FOR LEGHORNS ON A CONNECTICUT POULTRY PLANT

broody coop in which broody birds are imprisoned to break them up. The inside walls of this building are ceiled with dressed pine, tongued and grooved stuff, which gives an air-space between ceiling and outer wall and ceiling and roof. Twenty feet of length of the east end is partitioned off for a feed cookroom, where all the mixed feed is prepared, and here is set the hot water heater, from which flow and return pipes extend the entire length of the house, below the roost platforms, to warm the house in very cold weather. It is the intention to keep the temperature from falling below freezing. A low fire is run in the heater all winter, it being kept just alive in moderate weather and quickened when a cold wave comes on.

There are yards eighty feet long by ten feet wide for each pen on both front and back of this house. The north yards are grass runs for summer and the south yards for exercise in early spring and in the later fall, when it is too cold to have the slides open to the north. A new poultry house is being built directly south of this one, about one hundred and twenty feet distant from it, and it is the intention to foreshorten the south yards of this old house to thirty feet, giving the space thus rescued to the yards of the new house, which will be on the north side only

HOW THE GREAT LAYERS WERE FED

There were twenty-two birds, twenty females and two males, in each pen, and they were fed a mash every day; in the afternoon. In the morning, early, each pen received one pint of good sound wheat, which in winter was scattered in the deep litter to start them scratching. About 9:30 a. m. one-half pint of oats was scattered in the litter of each pen. About 1 p. m. one-half pint of cracked corn was scattered in the litter. At 3 p. m. in winter and 4 to 5 p. m. in summer they were fed (in troughs) all the mash they would eat up clean in half an hour. The mash was made of mixed meals, mixed in the follow proportions: 200 pounds wheat bran, 100 pounds corn meal, 100 pounds wheat middlings, 100 pounds linseed meal, 100 pounds beef scraps. One third of the bulk of the mash was of clover leaves and heads secured from the feeding floor of the cattle barn, or clover rowen (second crop clover) cut into one-fourth inch lengths in a clover cutter. The clover was thoroughly soaked in hot water, and the mixed meal stirred in till the mash was quite dry and crumbly. Cracked bone (which is steamed bone dried and cracked), crushed

42—WHITE PLYMOUTH ROCK PULLETS

oyster shells, clean, sharp grit and fresh water were before the birds all the time. In winter two large mangel beets were put in each pen daily, and in summer green food in plenty was fed. This last consisted of winter rye, which is the earliest accessible green food, then clover and grasses, then rape, the leaves of which are pulled while young and which yields four or five crops (or pullings). These were all run through the cutter and cut to one-fourth inch lengths. Prof. Gowell did not feed cabbage at all, having a poor opinion of it as a green food. The foods named are better, and there was an abundance of them.

It was a great pleasure to visit the Maine Station and see the breeding-laying stock there, especially the about 700 head of pullets raised this year. The visit was on the last day of October, the end of the laying year, and the old birds were being removed preparatory to bringing in the next year's layers. We personally handled the birds shown in the frontispiece of this book, and were surprised to find the hen that had laid but eight eggs in the year was one of the finest looking birds there. We carried her out to the pen of birds which were being sold to market, and confidently assert that the forty to fifty birds being put out there because they were indifferent layers were a better looking average lot of hens than the pen in which Nos. 617, 318 and 1003 (see frontispiece) were put.

Any poultryman would have said the same, and a man going there to buy a pen of birds and buying "on their looks" would be certain to take that pen of culls. The inestimable value of trap nest selection of laying-breeding stock is clearly shown in such a lesson as we had that day!

NOTE—The dry-feeding method is now being tested at the college with so good results that those in charge are enthusiastic about it. The great saving seems to be in the item of labor. The food the fowls consume costs about as much, but one man can tend 1500 birds with comparative ease. So far the results seem to prove that the germs in the eggs are stronger and consequently the eggs hatch better and the mortality is not so great among the young chicks. If experiments extending through a series of years prove this to be true, the days of mash feeding are numbered.—EDITOR.

MINORCAS AS EGG PRODUCERS

FOR ANNUAL EGG PRODUCTION AND SIZE OF EGGS THE MINORCA IS UNEXCELLED—STANDARD-BREDS, WINNERS AND LAYERS—EGGS WEIGHING TWO POUNDS PER DOZEN

J. H. DOANE

THIS article is based on the value of Black Minorcas as egg producers. It is generally considered poor policy to claim supremacy for any breed of poultry in a general sense, but when it comes to egg production it does not require any stretch of facts, or illusionary views, to assert that Black Minorcas are the acme of perfection as egg producers. It is a fact undisputed by those who have given them a trial that Black Minorcas are not outclassed in number of eggs laid by any other breed or variety, while in size of the egg Black Minorcas are in a class by themselves. We have not had experience with all breeds by any means, but we have bred several of the leading varieties and by every test the Black Minorcas have proved themselves the best layers at all seasons. Four years ago we decided to breed them exclusively.

When one attempts to show that the size of the egg should affect the price it is invariably cited that eggs are sold by the dozen. This claim is granted; yet the housekeeper chooses the big ones every time. Any other product of the farm—horses, cattle, cheese, butter or grain, has a market value which fluctuates with the quality. A high stepping cobby horse built with the curves and symmetrical points of equine beauty, or an extra good-looking draft horse, to say nothing of superior track horses, will command prices far in advance of the market. This same applies all through the long list of farm products and there is no exception in the real value of eggs. To say that an egg is an egg is about as unmeaning as the assertion that a hen is a hen. The cost of keeping a cow that produces forty pounds of milk per day is no more than that of keeping one that will produce but twenty pounds, and the cost of keeping a hen that will yield two hundred eggs in a year is no

41—ROOSTS AND DROPPINGS BOARD

more than that of keeping one that will lay but half that number. The care in either case is equal, and the pleasure and profit is all on one side.

Black Minorcas will produce as many eggs in a given time with proper care as their nearest competitors—the Leghorns—while their eggs will outweigh the other breed by several ounces to the dozen. Again, Minorca eggs do not have to be sold at the same price as other eggs, for their large size will readily command a higher price. We have no trouble getting three to five cents per dozen above market price for all the eggs we can spare, and other Minorca breeders realize about the same additional profits. As well charge market price for small potatoes as for small, under-sized eggs; and if Minorca breeders could have their way eggs would be sold by weight instead of by the dozen. This would be but justice due the purchaser, and such a measure would be a step out of dark age customs.

Although evidence is unnecessary, there appears on page 10 an illustration which is a half-tone from photo of one dozen eggs, every one of them laid by Black Minorca pullets, showing an exact net weight of thirty-two ounces—two pounds—for one dozen of pullet eggs. In evidence of the fact that standard-breds are the best layers I may say that six of the eggs were laid by the first-prize pullet at Madison Square Garden—the Crystal Palace show of America. Those six eggs were laid in six consecutive days—March 22, 23, 24, 25, 26 and 27. Four of the other six were laid by one of the two pullets in the second prize pen at the same show and the other two by an unshown pullet of same breeding and quality. Can you produce pullets of any other breed that will equal that weight? Note the even size and pure white shells, smooth as you could wish; the old saying, "Handsome is as handsome does," is illustrated in a clear, practical and valuable manner here.

These pullets were hatched in May and June and commenced to lay in November and December. The first-prize pullet began to lay eighteen days before the New York show (which fact disproves the worn-out theory that a laying pullet loses her bloom), and laid during the exhibition. From March 1 to 30, inclusive, she laid twenty-four eggs. We have three other pullets that have made even better records than this one, and it is not necessary to point out this particular pullet except to demonstrate that to breed to the highest type of perfection does not detract from the practical value of the Black Minorcas as egg producers.

We have bred Black Minorcas for sixteen years, and while our chief aim has been to breed near to the standard, we have also worked to build up an excellent egg producing strain. As the chief object in the life of a Minorca is to keep the egg basket well filled, it would not be possible to get a better variety to work upon successfully. For years we worked at a disadvantage, and in the light of modern record keeping or pedigree breeding, we worked in the dark. It is safe to assert that no man can accomplish the best results in breeding either for standard requirements or in producing an excellent egg strain without the use of trap nests. We have our pens fully equipped with them and we are convinced without a possible doubt that perfection in pedigree must come through their use. There is one exception, and that is separate matings of one male and one female, but that of course is not practical, nor profitable, nor will it ever come into general use, while the trap nest is sure to come into general use, or some will surely learn that they are not up-to-date. The expense is slight and the profits derived are very evident. The record sheet shows that two pullets have not paid their board, while one has laid but six eggs in five weeks. To be sure that does not read as nicely as the rest, but facts are appreciated and records are helpful; if all were extra good layers the trap nest would have no value in determining the best.

Black Minorcas are conceded to be one of the hardiest varieties we have among all the standard breeds. "A chick well hatched is half raised" will apply to them if to any breed. The chicks are quick growers; the cockerels are early crowers and the pullets lay early and often.

The soil around the buildings is of a sandy nature and well drained, but the greater part of the farm is composed of a heavier soil and is very productive. There is no way to grow chickens like turning them out to grass and giving them free range. The fields around my buildings are black with birds and they come to the house only when it is too dark to catch any more crickets.

It is a well known fact that fowls grown in the northern part of the state have a luster of plumage hard to equal where the climate is more mild.

If any one has any doubts about Black Minorcas being the "Acme of Perfection" as winter layers, we invite you to give them a trial, for we are certain you have never done so. The winters of northern New York are severe in the extreme, and still we have invariably had a good supply of fresh eggs, and the keynote to it all is exercise.

Keep the hens busy hunting for grains in deep litter and an increased egg yield will be the result. Eggs will hatch better, hens will thrive better and grow larger. It is the plan of nature, as it promotes vigor. To give your birds proper attention requires a love for the business and continual observation of details of the business.

BETTER LAYERS AND MORE OF THEM

INDIVIDUAL RECORDS THE MEANS BY WHICH A STRAIN OF LAYERS IS PRODUCED—
PROPER FOOD, SUFFICIENT EXERCISE AND FREE RANGE—CARE OF YOUNG—AN EGG
PRODUCING RATION—TRAP NESTS NECESSARY—NEW LAYING AND BREEDING HOUSE

C. BRICAULT, M. D. V.

WHEN some nine years ago we began poultry keeping it was with the fixed purpose of breeding a strain that would be a source of profit as egg layers. We did not, however, lose sight of the profits to be derived from dressed poultry, as about fifty per cent of the product would be males. In order that they should be profitably disposed of they must be of a quick maturing breed, possessing the desired qualities for our markets, namely, yellow skin and legs, and round, full breasts.

We had these fixed ideas about what we would like in a breed, and also what we would like this breed to produce in the number of eggs laid. How we were to accomplish all this seemed to us easy at that time, but as difficulties began to present themselves we began to discover that our task was not so easy as it seemed at first. We had been reading for some time about the good practical qualities of some families of White Wyandottes, so we decided to adopt that variety for our purpose. We looked up those breeders having in their birds the good qualities that we wanted and bought eggs and stock.

Almost from the start we saw the necessity of knowing just what each bird was doing in order that we could work along the lines we had mapped out. While we were devising some means by which we could record each hen's doings, we spoke to a friend of our plan and he offered to make a nest box which would solve that question for us. I well remember coming home that evening with the box under my buggy seat and my imagination afire anticipating the wonderful results that would follow. The box was placed that night in the pen and early the following morning I was on hand to watch its operation. Although that box was not all that could be desired, it opened to us the great possibilities to be attained by its use. Since then we have used and are using different kinds with varying success, and we would not think we were doing our best without trap nests. It is only by keeping an accurate record of each hen that we can make any progress in breeding. We mean any real progress. Don't tell us that it is by breeding from a whole pen that you can build a strain of layers. That a whole flock of hens can be bred to lay 200 eggs or more a year is no more to be doubted, but the shortest road to that end is by systematic, careful line breeding of males and females, and by knowing just what each hen is doing.

We believe in line breeding as the most successful method to obtain results, whether breeding for feather or for eggs and meat. Many things that happened to us in our breeding operations justify us in believing that feathers and eggs can be bred in one individual—or, in other words, we are trying hard to breed 96-pointers and 200-eggers in one individual. We have not as yet made much progress in the 96-point line, but we have made some; and what little we have gained makes us firmer in the belief that it can be done. One thing we have proved to our satisfaction is that a bird need not be brassy to be vigorous, as some would have us believe. We have birds that are chalk white and still are just as vigorous and whose eggs have proved just as fertile as the most brassy ones. Brassiness has simply got to be bred out, just as large egg yield has to be bred in. We firmly believe that large egg yields can only be had when we breed in line from our best layers.

44—HOUSE RECOMMENDED FOR SMALL FLOCK

This breeding in line means to transmit the qualities and organic development of our birds by using a system with which the most successful breeders of animals in the world have been successful. It means also that the egg must be well hatched and the chick well brooded, so that it begins to develop well from the start. The proper development of the pullet is one of the principal factors which go to make the great layer. Proper food, sufficient exercise and free range when old enough are three requisites of a successful method of growing the future layer.

HATCHING AND BROODING

We do all our hatching with incubators. The incubators are placed in our house cellar, a part of which has been partitioned off for that purpose. A large window on the south side admits all the fresh air that is needed. We have had some grand hatches from our machines placed in this cellar. The eggs are put in the incubator in the morning, and at night the machines are generally running steadily. Before putting the eggs in we light the lamps and get the temperature to 102½. When it has been running a whole day steadily we put the eggs in. That, of course, reduces the temperature. We fill the incubator in the morning so that at night the temperature is running at the proper degree of heat, and can be left to itself without any fear. We turn the eggs twice a day. On the sixth day we make our first test; all those proving infertile are discarded. On the fourteenth day we make another test and all those not showing strong germs are rejected. On the eighteenth day we place all the eggs from each hen under a small wire box in the tray. This is done in order to identify each chick when hatched. After the chicks are thirty-six hours old we take the contents of each box, band them, inscribe the number on our record book, and place about sixty in each brooder. For young chicks we use pigeon bands, which are small aluminum bands numbered as desired.

Before taking the chicks from the incubator we heat the brooders to 90 degrees. The bottom of the brooders is carpeted with green cut clover, into which we throw a small handful of millet seed. We always use great care' in taking the chicks from the incubator, so that they will not get chilled. For this purpose we have a basket which has a lid, and the whole inside is lined with two inches of cotton batting, and is previously warmed before putting the chicks in. We believe more chicks die through not being properly cared for at the time they are moved from the incubator to the brooder than one would believe. Those chicks that paste up behind in a few days have been chilled during this removal. In the brooder we place a small fountain filled with lukewarm water, a box containing grit, charcoal and bran. Our brooders are placed inside a small house, six feet wide and eight feet long, having a window and door on the south front.

FOOD FOR YOUNG CHICKS

Our food for young chicks is a prepared bread, composed of two-thirds whole wheat flour and one-third cornmeal, mixed with milk and baked thoroughly. When dry it is crumbled and fed on a small piece of board. They get nothing but this bread and the few millet seeds that they scratch for in the litter for the first fifteen days. We have tried all the different feeding methods of which we have read and heard, but this one has given better results than any other. When two weeks old we add a little cracked wheat and pin-head oatmeal to the millet seed, which we scatter in the litter as before, still continuing the bread every three hours as before. We feed five times a day during the first two weeks; four times a day until eight weeks old, then three times a day for breeding stock.

When two weeks old we reduce the heat in the brooder to 85 degrees; at one month old to 75 degrees, then gradually to 70, until they are well feathered out. At four weeks old we use whole grain and begin to feed a little cracked corn. We aim to give our chicks exercise from the first day. The floor of the brooder house is coarse sand. This we also cover with green cut clover, and into this we throw all the grain. They soon learn to scratch and the way they make the clover fly is amusing. The more they scratch the healthier they will be. After the first two weeks we open the sliding windows to the yards and let them out. If any snow covers the ground we sweep it away for a few feet and let them go. We continue this way of feeding until six weeks old, when we discontinue the bread and replace it with a mash (given at night) composed of one-third cornmeal, two-thirds bran, with about five per cent of animal meal scalded. Three times a week we begin to feed a little cut bone, and what a sight it is to see the little fellows tumble over each other for these bits of bone. For green food before grass begins to grow we use cabbage leaves, cut up fine, also lettuce leaves.

As soon as we find that the chicks prefer to roost on the brooder rather than to stay inside at night, we remove the brooder and place at the back of the house two small roosts. We then change the clover that has been used for scratching and replace it with clean material. When the chicks are large enough so that we can distinguish the sex, we separate them, placing the pullets on one part of the farm and the cockerels on another, giving each free range, a privilege which they do not fail to use. About that time we begin to feed three times a day, the same as we do the breeding stock. The band used on the chicks when taken from the incubator, being only a flat ring of aluminum, can be enlarged to suit the growing size of the chick's leg until now, when we remove this band and replace it with a permanent one, always giving to each chick the same number. They are weighed and all the details entered in our book. We weigh them again on the 1st of October, before placing them in the breeding pens.

EGG PRODUCING RATION

The question of what to feed for best results has kept us awake many a night. After feeding a very narrow ration and also a very wide ration, we came to the conclusion, by observing the results of both methods, that a ration analyzing 1:4 gave us the best results with Wyandottes. We are firm believers in as large a variety of foods as possible for best results. Green food in winter must be given with a liberal hand. Animal food is absolutely necessary, for you can not get eggs without it. We believe in corn, but not corn alone. Cut clover we believe in, and should not think of going without it. We want to feed our hens with as cheap a ration as possible consistent with the largest egg yield. No doubt our method of feeding can be improved on, and we should much like to do so. We buy the best grains on the market, and all the hard grain is fed in litter six inches deep. For grain we use wheat, oats, barley, corn and buckwheat.

We have adopted feeding the mash at night, and while we doubt if this manner of feeding has improved the egg yield, we continue to do so, first, because it has not given us worse results, also because it cheapens the ration, and because it is much less work. Our mash at this season of the year is com-

45—YOUNG BARRED ROCKS AND THEIR COLONY-HOUSE

BREEDING FOR EGGS

posed as follows: Bran, 20 pounds; ground oats, 20 pounds; ground barley, 15 pounds; cornmeal, 15 pounds; buckwheat middlings, 10 pounds; linseed meal, 10 pounds; meat meal, 10 pounds. We prepare the mash by boiling about one-quarter the quantity of clover to what we use of the ground grains. To this we add the animal meal, about one-eighth to one-quarter of an ounce to each hen. When this is boiling we add the meals. We stir the whole to the proper consistency. We like the mash to be just a little "sticky."

Our bill of fare is arranged as follows:

	MORNING	NOON	NIGHT
Monday	Wheat.	Cabbage, Oats.	Mash.
Tuesday	Barley.	Mangels, Cut Bone.	Mash.
Wednesday	Corn.	Cabbage, Wheat.	Mash.
Thursday	Oats.	Boiled Clover.	Buckwheat.
Friday	Cut Bone.	Mangels, Barley.	Mash.
Saturday	Wheat.	Cabbage, Oats.	Mash.
Sunday	Barley.	Cut Bone.	Wheat.

Three days in the week we use vegetables, cut up in small pieces and boiled, for the mash instead of clover. On the days we feed cut bone we feed only one-eighth of an ounce of meat meal per hen and all the cut bone they will eat.

In each pen is a large trough in which the mash is fed. This trough is long enough to permit of all the birds eating at one time. Every month we weigh our birds and any one that is found overweight is put in a spare pen and dieted for a few days. When only the male bird is found too heavy we close him up in the laying room while the hens are eating the mash, and allow him to eat only the hard grain in the litter. All our pens are piped and the stock has constant access to clean, fresh water, cool in warm weather and warmed in cold weather.

If at any meal the hens do not seem to be hungry, we make a note of that pen and feed only a small quantity at the next meal. If they still seem to lack their usual appetite we omit one meal. We aim to be very regular in the time of feeding; in fact, we believe this small point is very important. By properly feeding our hens we can obtain results which a less careful poultryman will not, but no matter how well we understand this difficult part of the work, and no matter how careful we are in making use of our knowledge in this direction, it will be observed that in every flock of pullets raised, fed and cared for in the same manner and under the same conditions, several will lay almost double the number of eggs that others will in the same pen. With but few exceptions, these large egg producers were born with this valuable trait, and it is by breeding from those heavy layers that we can increase the average egg yield of our flock. Breeding systematically and persistently from our heaviest layers will develop a strain of layers which will pay us generously for our work. No other method will give us as good results.

TRAP NESTS NECESSARY

In order to follow this system of breeding it will be found of absolute necessity to ascertain the number of eggs each hen lays. Fortunately, we have at our disposition the automatic nest, which will help us to accomplish this accurately. This valuable addition to the practical poultryman's needs has been severely criticised by some, but its advantages can not be overlooked and no real progress can be made without its use. By placing these trap nests in the pens it will be an easy matter to distinguish the best layers from the poorest. The members of the flock that do not lay enough eggs to pay a profit should be disposed of to the butcher.

You can overcome the greatest drawbacks to trap nests by putting in the pens half as many nests as there are layers and placing them on a platform twenty inches above the floor. By this arrangement the work is reduced to a minimum and the hens have the advantage of using the whole floor space. By this plan you can gather the eggs with more ease and at a saving of fifty per cent in time, as compared with nests that are under the drop boards. With this number of nests in the pens there will be no need of visiting them oftener than four times a day.

KEEPING THE RECORDS

When your pens are equipped as advised above, the next thing to do is to place a leg band around one leg of each hen.

46—BARRED ROCK AND WHITE LEGHORN FUTURE LAYERS

These leg bands can be stamped with a number, letter or both; then you are ready to begin record keeping. As you go the rounds of the nests, you release the hens confined in them, note their numbers on the leg bands, and mark each egg, or enter it at once on record sheets kept for the purpose. You may find some customers who object to having anything written on the eggs they buy; many grocers and merchants object to this. The only remedy is to enter the numbers on the egg record sheet, or on a small slate carried around by the attendant, and, later, enter them on the egg record sheets. If the eggs are wanted for hatching, it will then become necessary to mark each egg.

There is no need of complicated record keeping. The record sheets on which are written the number of each hen and the dates on which she lays, checked off, and a small book are all that are required. Every page of this book is ruled off in three sections, and in each of these is entered the number of one hen, her record, her dam's record, and her sire's dam's record; also the number of each chick hatched from her eggs. When a chick dies, the letter "D" is written across its number. When we are ready to begin hatching, each hen's eggs are incubated separately under a hen or placed in a compartment of a pedigree tray in the incubator. When the chicks are hatched each one is marked by placing around one of its legs a small leg band. As they grow older these bands are changed for larger ones.

MATING FOR EGG PRODUCTION

One of the most important points in mating your pen for egg production, is the selection of male birds, for we lean to the belief that it is through her sons that a great layer transmits her qualities. Use none but well developed, vigorous sons of your very best layers. Another equally important thing in selecting your breeders, is vigor. Choose only the most vigorous hens and cockerels. Vigor is the outward sign of a strong constitution, and a good layer must be strong and vigorous to enable her to digest and assimilate the food necessary to lay a large number of eggs. By thus selecting each year the most vigorous descendants of your best layers, you will intensify both these qualities in your strain and produce layers that will lay more eggs than their ancestors did.

When you have arrived at this period of your breeding operations—that is, when by a few years of this systematic breeding, you have fixed vigor and a good egg record in your strain, you can profitably practice some inbreeding. By inbreeding you can improve your egg yield quicker than by selection. But be very careful to select only the most vigorous and healthiest individuals from your few best layers.

We will suppose you have just such a hen with a large egg record. You can mate her to her most vigorous and best developed son, and in this same pen you may also put the hens which have given you the largest number of eggs. The following year you can mate one of these inbred cockerels to the pullets bred by the first cockerel, but out of the other hens —they will be half-sisters. Of course, you will always pick out your highest record hens to mate to these choice breeders.

You will be pleased with egg records of the pullets bred from this last mating. The majority will be excellent layers, and will be the very best of breeders. If you have been careful to select only vigorous birds and the best layers in your matings, you will notice very soon a great improvement in the average egg yield. You can then use these inbred cockerels on unrelated hens, but only on the good layers, and then follow up as before.

Great productiveness in our hens is a trait which can be easily fixed by breeding. The principles governing our breeding are the same as those which apply to all other classes of animal breeding; it is only the application that differs. With the fancier it is feathers; with us it is eggs; both can be developed to perfection by the same principles of breeding. There is nothing to prevent you, if you so desire, from improving both the egg yield and exhibition points of your strain, but progress will be much slower. The results, however, will be more pleasing in the end. To us this question of breeding layers is a most fascinating one, and it is one which offers more real advantages to the interested poultryman than would be believed at first. Breeding from our best layers systematically is, to our way of thinking, the only sure way of increasing the profits.

NEW STYLE LAYING-BREEDING HOUSE

Several years' use of the "open scratching-shed" house made us appreciate the full value of fresh air as an important factor in keeping our breeding stock in the best of health, resulting in more fertile eggs and in stronger chicks, more easily raised. With all its many good points, this style of house has many disadvantages, especially for our cold climate, which one would like to have eliminated. We were also familiar with the closed house and knew its many faults. Hence, in building our new house we knew what to avoid and what improvements to make. The result was a house which we believe to be the best one yet devised in which to keep our breeding-laying stock.

We are firm believers in pure air for our hens. We have had ample opportunity to note its good effect on egg production, and it is our opinion that much of the failure in keeping hens profitably is due to lack of pure air in the poultry house. But strong as we are in our belief of the necessity of fresh air we do not believe that our hens should be exposed to the discomfort of a cold roosting room. That is why we close the doors of our house after sundown during the cold weather and thereby provide them a warm room in which to sleep. During the day, when they are scratching in the litter, they do not mind it at all. We noticed that all the hens would stay in our "open shed" practically all the time during the day, even when it was very cold, so that the roosting room was practically lost room.

The large door in the south front was the simple solution to this problem. With correct handling, this door allows us to give the hens all the pure air possible, during the day, and when the house has been so ventilated in daytime, it contains all the pure air needed from the time it is closed at night until it is opened the next morning. Of course during mild weather the upper part is left open and the cloth curtain

47—CONVENIENT COLONY HOUSES

BREEDING FOR EGGS

inserted in the opening for the night; the same thing is done during a very stormy day. On cold, bright days this curtain is omitted. If not too windy on these cold days the whole door is opened, and the sun just covers the whole floor of the house. How the hens like this can only be appreciated by going from pen to pen and noting their contented appearance as they stretch out in all manners all over the pen.

The pens, which are 10 by 12, will easily accommodate twelve to fifteen females and one male. It will cost less than the open-shed house, and no more than the closed house, over which it has so many advantages that those using the closed house would never return to that style if they once tried this one, for it combines all the essentials necessary to a successful poultry house.

DIMENSIONS OF THE HOUSE

Our house is 10 feet wide, 7 feet 6 inches high in front and 5 feet 6 inches in rear, divided into pens 10 by 12. The frame is 2 by 4 for sills, plates and rafters, and 2 by 3 for studding. The sills are laid 10 inches from the ground on cedar posts sunk 2 feet in the ground. It is boarded up with hemlock boards, first by imbedding the bottom board 4 inches in the ground and having this bottom board come to cover half the width of the sill, or 1 inch only. Then the upper boarding begins at this point, being nailed on the other half of the sill, thus making a tight joint at the bottom and avoiding any drafts—a small point, but an important one.

The house is filled inside with gravelly sand up to the top of sills, making a dry, warm floor. To make the house tight along the back wall near the roof, the rafters are cut even with the plates, and boards run up even with top of the rafters. Then in laying the paper we begin at the back, lapping the paper well on the roof boards. Then a course of shingles is nailed on this, and another on this one to break joints. Then the roof paper is laid over the shingles, letting these project about 6 inches from the back wall, to allow them to carry the water away from the building. The paper on the back wall is laid vertically, and lapped about 3 inches; the roof paper is laid lengthwise of the building and also well lapped.

We first cover the boards with a cheap grade of building paper; over this a thickness of Cabot's quilting is laid, then over all the Neponset Red Rope Roofing, the whole held in place with small nails and tin heads, which come with the paper. These tin heads should be painted, when they will last for a good many years. We prefer to nail the paper in this manner than to use cleats, which we do not think as good.

THE SOUTH FRONT

The south front contains a large door, 4 feet 6 inches by 6 feet 6 inches; also a window on each side of it. This door is divided in two parts, the upper part, which is 2 feet 6 inches by 4 feet 6 inches, is hinged to the plate and swings up against the rafters, to which it is secured by two hooks. On stormy days, or on mild nights, this upper part of the door is opened and a cloth curtain is inserted in its place. The lower part of the door is hinged to the side and opens sideways. Behind this lower part of the door is a slot-like arrangement, into which is slipped the curtain which is used in the upper part when this is opened. This cloth curtain is simply a 2-inch frame, on which is tacked oiled muslin. The windows on each side of the door are ordinary 2-sash 12-light windows, and can be operated up and down at will. In warm weather, when the doors all open, also the windows, this house is practically an open shed, and the most comfortable place possible for the hens. A small opening, one foot square, was placed near the floor, next to the door, to allow the hens the use of the yards when it is necessary to close the bottom of the door in early fall or during rainy weather. This renders it optional with the fowls to go out or stay within.

INSIDE ARRANGEMENTS

The divisions between pens are made solid, except the doors, which are 2-inch wire netting nailed to 4-inch frames. A good fitting door we appreciate very much, and ours were

48—VIEW ON EXTENSIVE WHITE LEGHORN FARM

made to fit well. They are hinged on double-action spring hinges and work perfectly. Probably because our doors fit well and we use double-action hinges, we have never felt the need of a passageway. Going through the pens with a pail in each hand, it is practically no trouble to push the doors open with your foot. But in ninety per cent of the poultry houses we have visited we have found very poor fitting doors, and no wonder these poultrymen want a hallway in their houses.

Along the back wall, 18 inches from the floor, is placed the drop board; 6 inches above that two roosts, which are 2 by 3, with corners rounded off and laid on the 2-inch side. Sixteen inches above the roosts is a row of coops, 30 inches in depth, running the whole length of the back wall. These coops we find very handy for extra males, broody hens, or to put in a pair or trio preparatory to shipping.

In the division between pens are the drinking vessels, raised 10 inches from the floor. These are made of galvanized iron, 14 inches in diameter and 6 inches deep. Next to these is a self-feeding box, 36 inches long, divided into three compartments, one for beef scraps, which we always keep before our hens; the others contain oyster shell, grit and charcoal.

EGGS AND EGG FARMS

The floor of each pen is carpeted with clover hay, into which all the hard grain is fed. Taken altogether, this house as it is to-day is giving us the best of results, and is to our way of thinking the most comfortable, practical house we have ever used. It combines all the essentials of a practical, successful, up-to-date poultry house.

TRAP NESTS A NECESSITY

A HOME MADE TRAP NEST WHICH IS SAID TO DO SATISFACTORY WORK—A NOVICE CAN MAKE IT—ITS CONSTRUCTION WILL OCCUPY TIME WELL SPENT

CLARK AND TROLL

LIKE most breeders of the present day who have decided to use trap nests in their breeding pens, we, early in our experience with the fancy and market poultry business, came to the conclusion that the only practical way to breed a strain of birds up to the highest point of excellence, in the shortest possible time, either for show purposes or heavy egg production, would be to ascertain with absolute certainty the sire and dam of every bird we produced, whether it be a prize winner, an exceptionally fine layer, or both. As our desire was to build up a strain of heavy laying prize winners, we set about devising a nest box that would assist us in securing the information we needed, and as it has to be positive (no if's about it, and therefore no two hens getting on the same nest at one time) we studied the needs thoroughly, and finally succeeded in making the "Champion Trap Nest."

If all who may use this nest box shall be as successful as we have been we will be gratified indeed. After experiencing the great advantage we have gained by its use, we would not now go back to the old hit-or-miss way of breeding. We would just as soon think of going back to the breeding methods of twenty-five years ago.

As will be seen by the illustration herewith, we make three nests in one section. We need only one size, each nest being 12 inches square, inside measurement. To make one section this size, we use 1-inch pine boards, 12 inches wide, for the bottom, ends, divisions and doors. Cut one piece 38 inches long for the bottom, two pieces 12 by 13 inches for the ends, and two pieces 12 inches square for the center divisions. Cut seven laths, each 40 inches long, five of which are for the top, and two for the back, to hold the nesting material in.

First nail each of the two ends onto the outside ends of the bottom. This makes it 40 inches long. Then nail the five laths on top, and next the two on the back. Then put in the two center division boards, thereby forming three nests each 12 inches square. To make an alighting board, cut three pieces about 15 inches long by 1 and 1½ inches. These have to be nailed on the under side of the bottom board, one at each end, and one in the center. Let each one project in front of the nest box about 5 or 6 inches, and nail a lath onto the ends of each.

49—GUARD AND TRIGGER WIRES

Now comes the more difficult part of the work, though I believe we could make a whole section in less time than it takes to write this. Cut three pieces, each 12 inches square, for the doors, and from the center of each cut a circular piece 7 inches in diameter. Now you will need some No. 14 wire for Ill. 49. Cut two pieces each 11 inches long and with a pair of plyers bend the wire as shown to left of the illustration, the long stretch in the center of this wire is 8 inches, which leaves 1 and ½ inches on each side to bend into shape. With some three-quarter inch poultry netting staples fasten these wires on the door, one on each side of the opening at equal distance apart at top and bottom, having the lower ends about on a level with, or a trifle below the lower edge of opening. which will allow the curtain to be raised above the trigger which holds it up. Next, cut a piece of wire 16 inches long and bend it in the shape shown at the right of Ill. 49; then cut a piece 2 inches long and bend one end to form a loop by which it will hang upon the center of this long wire. In the illustration it is shown caught up in position, but when released it simply hangs from the 2-inch center bend of the long wire.

Fasten these combined pieces to back of door with two staples for hinges, so as to swing free and easy. See to right of Ill. 49. Drive a staple about three-fourths the way into top of

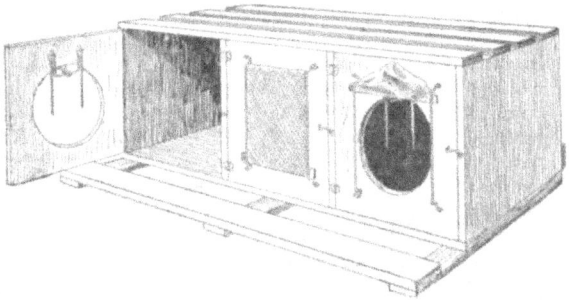

50—CHAMPION TRAP NEST

the circular cut, as shown in same. It may also be seen on Ill. 50. This forms a rest for the trigger (the 2-inch piece of wire) when it is placed in position to support the curtain.

The hen on entering the nest pushes the 16-inch wire and the trigger, until the latter drops from its rest and releases the curtain, which then falls and covers the opening.

To make the curtain cut a piece of muslin 8 inches square, and tack it upon a piece of lath 1 by 9 inches, which leaves ½ inch extending on each end, past the curtain. Slip the laths under the guard wires to be seen in Ill. 49, and also shown on the door of the center nest in Ill. 50, and after tacking the curtain at the top, above the opening or doorway, drive a staple into each end of the lath immediately outside the guard wires, so as to keep them in place when the hen may try to get out of the nest. Be careful not to drive the staples too close to the guard wires, or they will cause friction. Procure hinges and hooks for the doors and the nests will be ready for use. Make two wire hooks and fasten them to the back of the nest box. Place two staples at proper distance apart, in the wall of house, which forms the back of nests and hang the nests up to these; it is then out of the way.

After getting the nests in position, raise the curtain above the openings, slip the trigger through the staple, and let the curtain rest on it until a hen enters, when down it will come, leaving her a prisoner until released, and no other hen can possibly get in with her, nor can she get out until released.

So much for making the nest. We trust it has been explained clearly, so that all will understand how to make the separate parts, and place them together in their right position.

Now for the advantages to be gained by its use. If we were to ask you if standard-bred stock were as good as common scrubs you would very likely, and with good reason, too, consider that we knew nothing about the poultry busi-

56

BREEDING FOR EGGS

ness. If we ask, is carefully bred, pedigreed stock, of any kind or variety, better than the carelessly bred, hit or miss kind (especially in the matter of transmitting qualities to their progeny) your answer would be, "Certainly it is." It is an undisputed fact that some hens will lay a far greater number of eggs in a year than others, also that some hens are much better producers of prize winners than others. To take advantage of these facts every down-to-date breeder should avail himself or herself of the opportunity to secure a trap nest of some sort, one that will enable the breeder to know the good or bad qualities of every bird, especially of those in the breeding pen. We venture to say that the time is not for distant when every breeder of note, who wishes to be up with the times, will use a trap nest, or resort to some other means by which he can know the breeding qualities of each individual bird, which he may wish to breed from. If one has not the time to keep the nest boxes in use the year round, as it happened with us this season, they can be used during the time you are saving eggs for hatching, and after the hatching season is over the traps may be fastened up out of the way until another season comes around. On account of so much other work demanding our attention this summer we had to dispense with the use of our traps after finishing hatching, but used them all the time we were saving eggs, so that we now know the sire and dam of each chick raised and all the desirable qualities of each for several generations.

To be a live, wide-awake American poultry breeder, one must be progressive, and to progress is to advance. Get out of the old ruts, and take advantage of every opportunity that is within one's reach. Just as the old style machinery in nearly every known vocation has gone out of use to be replaced with new and modern inventions, so it is with poultry breeding. The old way was very good in its time, but the modern methods will be adopted in future to produce "better poultry and more of it."

FEEDING FOR EGGS

A PRACTICAL STUDY OF EGGS

NON-TECHNICAL PRESENTATION OF THE FORMATION AND PRODUCTION OF EGGS THAT CAN BE UNDERSTOOD BY ALL—CAUSE OF DOUBLE-YOLKED, SOFT-SHELLED AND CROOKED EGGS, ONE EGG WITHIN ANOTHER AND ROTTEN NEW-LAID EGGS

G. BRADSHAW

ALTHOUGH eggs are a common article of food there is not a general knowledge amongst poultrymen as to their formation. The shell or envelope is white or colored according to the breed which produces it, and is composed of corbonate of lime, phosphate of lime, and animal gluten, salts of lime causing the particles to adhere. Soft eggs are either eggs without a shell, or the shell may be so thin as to feel soft through the deficiency of salts of lime. It is a matter of surprise where a hen finds all the lime necessary, for if she lays 150 normal sized eggs in the year, she will have produced two pounds of pure carbonate of lime.

HENS ARE WONDERFUL CHALK MAKERS

Mr. P. L. Simmons, F. L. Z., on this subject in the Journal of the Society of Arts, says:—"If a farmer has a flock of 100 hens, they produce in egg shells about 137 pounds of chalk annually, and yet not a pound of the substance, or perhaps not an ounce, may be found on the farm. The materials for the manufacture are found in the food consumed, and in sand, pebbles, brickdust, pieces of bone, etc., which hens and other birds are continually picking from the earth. Their instinct is keen for these apparently innutritious and refractory substances, and they are devoured with as eager a relish as the cereal grains or insects."

If hens are confined to barns or outbuildings, it is obvious that the egg-producing machinery cannot be kept long in action, unless materials for the shell are supplied in ample abundance. If fowls are confined in a room and fed with any of the cereal grains, excluding all sand, dust or earthy matter, they will go on for a time, and lay eggs, each one having a perfect shell made up of the same calcareous elements.

THE SHELL IS A "SIEVE"

The shell is porous to such an extent that when examined under a microscope it has quite a sieve-like appearance, and is permeable by the air, otherwise the chicken could not live during the incubating period.

This porosity of the shell, although absolutely necessary when the eggs are to be incubated, is detrimental when they have to be used as an article of food, from the fact that by means of these minute perforations there is a continual evaporation, so that from the time the eggs are laid until consumed there is a wasting and deterioration of the contents, the extent of which is dependent on the temperature and other conditions under which they are kept, it being very well known that eggs deteriorate much quicker in summer than in winter.

FORMATION AND PRODUCTION OF AN EGG

Any one, upon opening after death the body of a hen, will find a cluster of eggs in formation much like a bunch of grapes and called the ovarium (see illustration). These, however, are but rudimentary eggs. I have counted as many as seventy in one bunch, and are in size from a pin's head to the full-sized yolk of an egg. Each of these eggs is contained within a thin transparent sac and attached by a narrow pipe or stem to the ovary. During the laying period of the hen these eggs are maturing and thus keeping up the supply which she lays.

These rudimentary eggs have neither shell nor white, consisting wholly of yolk, on which floats the germ of the future chicken; and as they become larger and larger they arrive at a certain stage when, by their own volition, weight or other cause, they become individually detached from the bunch and fall into a sort of funnel leading into a pipe or passage called the oviduct—this organ in the hen being from 22 to 26 inches long.

THE COATING OF ALBUMEN

During the passage of this egg or ovum to the outer world it becomes coated with successive layers of albumen—the white—which is secreted from the blood vessels of the oviduct in the form of a thick glairy fluid, and is prevented from mixing with the yolk by the membrane or sac which surrounded it before it became detached from the cluster. It is also strengthened by a second and stronger membrane, formed around the first immediately after falling into the funnel, and having what is like two twisted cords of a more dense albuminous character, called by anatomists chalazes, which pass quite

FEEDING FOR EGGS

through the white at the ends, and being, as it were, embedded therein, thus preventing the yolk and germ from rolling about when the egg is moved, and serving to keep the germ uppermost, so that it may best receive the heat imparted during incubation.

as double-yolked eggs, but more properly it is a double egg, the white being duplicated as well as the yolk. Should these yolks be fertilized and the egg hatched, we get the occasional four-legged or other chicken monstrosities.

A further result of stimulating food is varied from the

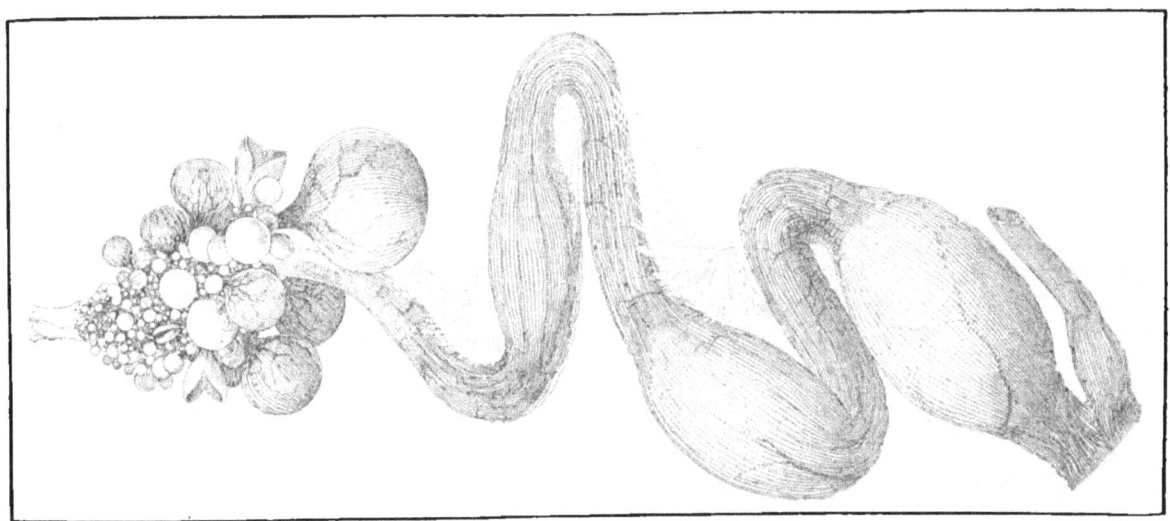

51—OVARY AND OVIDUCT OF A HEN

THE AIR CHAMBER

It is during the passage of the egg through the lower part of the oviduct that it gets covered with the two skins which are found inside the shell. These, although lying close around the egg, at the thick end become separate, and form what is called the air bubble or chamber. This, in newly-laid eggs, is a mere speck and is that portion which shows the result of the evaporation previously referred to. This speck of air space becomes daily larger as the egg gets older, and is frequently equal to one-fourth of the entire egg. This egg-chamber, if perforated with the finest needle, will prevent the egg hatching.

When the egg has advanced more than half-way down the oviduct, it is still destitute of shell, which begins to be formed by a process of secretion, and when about completed the various shades of brown and tinted coloring matter are imparted in those breeds in which colored eggs are peculiar; sometimes in very brown eggs white spots appear, but these can readily be rubbed off. When the shell and coloring are complete the egg continues to advance along the oviduct till the hen goes to the nest and lays it.

RESULTS OF TOO STIMULATING OR EXCESSIVE FOOD—DOUBLE-YOLKED EGGS

Eggs are produced from the surplus food, which is that over and above what is required for the sustenance of the hen, and if such is too stimulating or given in excessive quantities, the result is that in the former case the ova are produced so rapidly that sometimes two of them drop into the oviduct together, which results in the eccentricities which frequently puzzle the poultry keeper. These ova travel together along the passage and receive the white separately, but become enveloped in one shell, and when laid are commonly known

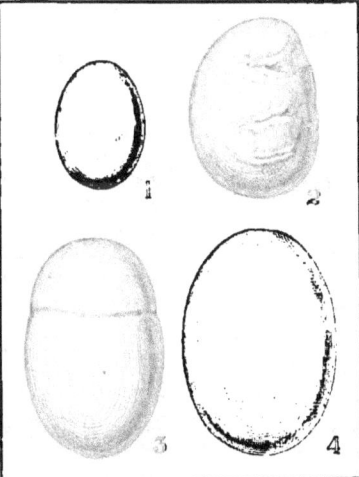

52—DEFORMED EGGS
1—"Marble-sized" egg containing albumen and shell, but no yolk. 2—Crooked egg. 3—Two eggs with imperfect shell. 4—Double-yolked egg.

above when the ova mature in excess of one a day. In place of falling into the passage in pairs, as above, the two drop in separately, but on the same day. This results in soft eggs, not from the want of shell-forming material, but rather because the shells can not be formed as fast as the mature egg is ready for such covering.

Crooked eggs are no uncommon thing in the poultry yard and are attributable as follows: Twenty-four hours are usually sufficient for the formation of a perfect shell, but when by stimulation a second ovum falls close on its predecessor, reaching it before laid, the second egg, which is up to this time soft and is lying against the hard one, becomes covered with a shell, and when laid presents a flat or crooked side, the result of its position against the hard one.

ONE EGG WITHIN ANOTHER — "MARBLE-SIZED" EGGS

To over-feeding is also attributable the further irregularity of one perfect egg being found within another, and caused by irritation of the oviduct, which contracting in front of the perfectly-formed egg instead of behind it, forces it back till it meets another yolk, when the two join and again become coated with the white and the shell, thus producing another wonder. Other abnormalities are occasionally seen, and particularly in the smaller poultry yards. Sometimes when the ova are nearly exhausted by continuous laying, the secreting organs may be most active, which results in small marble sized but perfect-looking eggs, which are merely a shell covering a portion of albumen. Such "eggs" when laid have the peculiarity of never having been at any stage attached to the ovary, but are a product only of the oviduct.

To the internal fatness of the hen are due other eccentricities than those mentioned, including the apparently paradoxical feat of laying rotten new-laid eggs, this being a not infrequent occurrence. The egg, being unable to force its way through the fatty oviduct, is retained two or three days near the mouth of this organ, and, if a fertilized one, the heat of the hens body tends to putrefy it, and when ultimately laid it is in an addled condition. To other causes, but principally diseased organs, is due a departure from the normal in the way of color. A hen which lays white or brown eggs, on rare occasions produces one almost black, while at other times these vagaries much resemble the dark green of the emu's eggs, and, in most instances, the shells are rough, wavy, corrugated, or otherwise irregular. Then there are instances of foreign matter being found in eggs, clots of blood being noth-

53—MISSHAPED EGGS UNFIT FOR HATCHING

ing unusual. This is the result of the breaking of a blood-vessel internally, and, again, possibly the effect of over feeding.

FOWLS PRODUCING FAULTY EGGS SHOULD BE DISCARDED

Fowls from whatever cause producing any of the above misshapen or otherwise faulty eggs should at once be got rid of, for although in some cases a reduced diet may bring them back to their normal production, still the slightest cause will frequently prompt the organs to their previous irregularities, the fowls thus becoming unprofitable members of the flock. As has been seen the majority of troubles mentioned are preventable ones, and largely due to the poultry keepers' mistaken kindness in overfeeding, but there are other ills of a more serious nature than those mentioned. The producing organs are of a most delicate nature, and, from the amount of work they have to do, are not only easily disarranged but are subject to a variety of diseases, the nature of many of them being unknown to the ordinary poultry keeper.

RATIONS AND FEEDING

ADVANTAGES TO BE DERIVED FROM FEEDING A RATION CONTAINING IN CORRECT PROPORTION FOODS WHICH PROMOTE EGG PRODUCTION—EXERCISE NOT ESSENTIAL

H. E. MOSS

ON THE general farm, where poultry raising is carried on as a department and not a specialty, there is usually a variety of crops grown which would furnish the bulk and the basis of a proper feed, providing feeding values were thoroughly known and understood. To these then there can be added by purchase the elements lacking and a correct ration devised which will be satisfactory in results as well as economical.

Of all the grains, corn, wheat and oats are the more readily available. To combine these with vegetable and animal food in order that the fowls may be funished with no excess of one element is the point to be studied and figured.

The feeding of a balanced ration to the poultry is now fully recognized as essential to success and profitable results just as on up-to-date dairy farms, the best of talent is employed and the feeding of dairy cows is reduced to a science. The principles involved are precisely the same.

Unless our hens will produce eggs during the winter months, when eggs are high, they become unprofitable. With eggs at from thirty to forty cents per dozen during November, December and January, each hen should earn at least one dollar above her feed, which she will if feed intelligently, and it will cost little, if any more, to feed her correctly than to feed at random or spend this most valuable season in experimenting. The majority of poultry keepers are too ready to blame the breed or seek some other remote causes when their birds fail to lay as expected. They are too willing to believe that the cause lies anywhere but in the feed, especially if they are feeding liberally. They consider a full crop as sufficiency. To illustrate the inconsistency of this, suppose we could induce a hen to consume three and one-quarter ounces of wheat a day; we know that two and three-quarter ounces of this is used in maintenance of the body, for fuel, energy and repairs. This leaves half an ounce to be converted into eggs, and this is assuming that all the food is assimilated. The extra half-ounce of wheat she has been given contains six-one-hundredths of an ounce of protein—and an egg contains about one-quarter of an ounce of protein. She has then available for each day just one-fourth enough protein for one egg, but the same one-half ounce of wheat contains one and one-half times as much carbo-hydrates and fat as one egg contains.

To feed wheat alone cannot result in greater egg production than is provided for by the protein that is supplied; at the same time there will be an accumulation of surplus fat. The bird is then said to be too fat to lay. The facts are that she has not stopped laying because she was too fat, but because she was fed a ration that prevented her from rapidly forming the albumen of the egg. There is no single grain that contains all the necessary elements in correct proportions or balanced to favor rapid egg production. It requires careful and judicious selection and mixing of a variety of palatable and nutritious grains in such proportions as to supply just the right amount of material for the specific purpose.

The question of a balanced ration is not as fully appreciated or understood as it should be, even by many who study and attempt to apply it. The very first question to be decided is, what shall the ration be? Shall it be 1:4—which is very narrow—or 1:4, 1:5 or 1:6—the latter being wide? As the figures to be adopted must vary with the climate and season as well as the activity or sluggishness of the bird, it

FEEDING FOR EGGS

will require a careful determination to decide this before attempting to formulate the ration. The thermometers at zero in Maine and 90 degrees at Jacksonville, Fla., call for an entirely different ration. In the former place it must be wide and in the latter narrow. The results of much careful investigation and study have demonstrated that a laying hen is enabled to do her best work with a ration with one of protein to four of carbo-hydrates where the temperture is uniform at 60 degrees. It is, therefore, necessary as it rises or falls to compensate for it in the feed.

The tables showing the analysis of feeding stuffs are so frequently published and so easily obtained that we shall not devote space here to print them, but shall show by example how the ratio of any formula can be determined:

FEED	PROTEIN	CARBO-HYDRATES	FAT
100 lbs. Cornmeal	9.2	68.7	3.8
200 lbs. Ground Oats	23.6	119.4	10.0
200 lbs. Wheat or Bran	30.8	107.8	8.0
200 lbs. Wheat Middlings	31.2	120.8	8.0
100 lbs. Beef Scraps	71.2	.3	13.7
	166.0	417.0	43.5
		100.0	2.3
		517.0	100.05

The oats, bran and middlings call for two hundred pounds in the ration, therefore we have multiplied the analysis given in the tables by two. When the fractional part of a hundred is used the same fractional part of the figures must be taken. In the above we add up each column, then multiply the fats by 2.3, as they are equivalent to them in carbo-hydrates, add the result to the carbo-hydrates, which gives us a total of 517.3. As the protein is the unit, we divide 166.0 into this amount and find our formula shows a nutritive value of 1:3.1. This is a very narrow ration, and as it contains one-eighth by weight of beef scraps, its fault is easily discovered and remedied by reducing the meat, which when fed in the mash should never exceed ten per cent of the grain by weight, and where bugs and worms are to be found on the range, five per cent or none at all is at times sufficient. But, remember that even if you can compound a ration showing the correct ratio without beef scraps or animal food in some form, you will not be able to raise your birds to maturity successfully without it. They must have a certain amount of protein that is derived from animal sources or they will not thrive or grow. They can not form feathers, frame or muscle without it. The following formula is an example of one that figures a safe ratio for general use, which is 1.4, but wrong as to components:

FEED	PROTEIN	CARBO-HYDRATES	FAT
100 lbs. Bran	15.4	53.9	4.0
100 lbs. Middlings	15.6	60.4	4.0
50 lbs. Cornmeal	4.6	34.4	1.9
25 lbs. Linseed Meal	8.2	8.8	1.9
	43.8	157.5	11.8

To use the above ration would be to invite defeat and loss, for several reasons. The first is, there is too large a proportion of bran, which contains much hard, indigestible fibre and would lead to irritation of the bowels, while a moderate quantity, say not to exceed one-fifth the grain, would be beneficial, as it acts as a mechanical stimulus, which is necessary. Another is the excess of linseed meal, which must be used very carefully, especially on birds that are to be raised to maturity. Another is the absence of animal food.

I believe the most abundant egg production can be secured during the winter months by keeping constantly before the birds after the first of November a supply of corn, wheat, oats, buckwheat and any other grains in separate hoppers, in addition to a hopper of beef scraps and an abundance of skim milk, with grit and green food, of course. They will then be under the pleasant surprise of an abundance constantly before them, be stimulated to early egg production and will consume just what they need of each article. This will work very well until spring, when it should be discontinued and systematic feeding again adopted; for as soon as the range affords feed, the amount and kind furnished should be regulated by the amount the range supplies.

The question of exercise will be raised by many and their opinions will be that where feed is constantly kept before the birds, they will spend much of their time in idleness, which they will, but which I also know is not necessarly inconsistent with steady egg production, as this is dependent on the material she is supplied with. Place a hen in a store box and

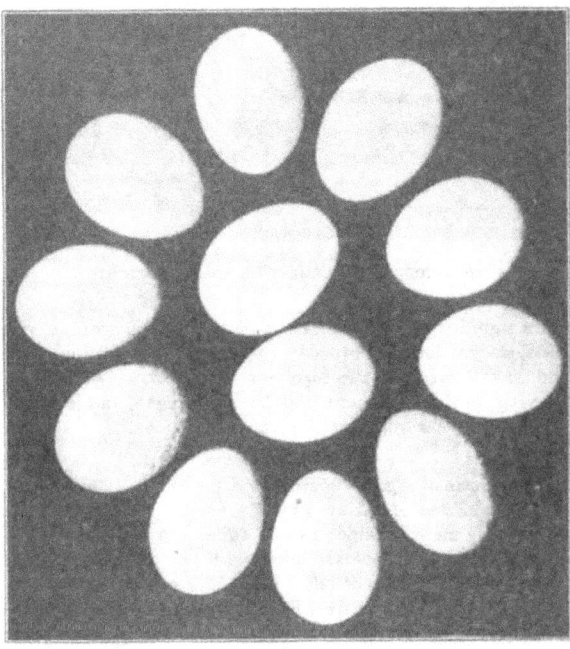

54—UNIFORM, WELL-SHAPED EGGS, SUITABLE FOR HATCHING

feed her on fat-forming food, and idleness will quickly put her out of the egg business; while with an egg-producing ration she can be kept laying continually while cooped. Of course the hatchable quality of eggs produced under such conditions is very low. This method is intended only where market eggs are desired, quantity and not vigorous, strong germs being aimed at, and the maximum quantity has in many instances been secured under precisely these conditions. There is little fear of over-feeding a laying hen, provided she is given enough albumen-forming material to work with and the fat-formers in only sufficient quantity to keep the engine running. The best egg-producers will always be found heavy in flesh (not fat) and you will never find a good layer in thin flesh unless she is exhausted and about to stop.

Where a maximum egg production is the first and most important consideration, as it is on the average farm, it will be conceded that the birds must be kept in perfect health and their physical condition maintained by a feeling of comfort. In addition to this, they must have the assurance of an abundance of egg-making material and secure a regular supply of it before they will undertake egg production. If their instincts teach them anything, it naturally is to delay undertaking the task of egg production until they are in good flesh and find themselves supplied with not only sufficient flesh-forming food to maintain this condition, but a surplus in sight to justify the attempt to convert it into eggs. A half-starved hen never has and never will be found laying. You say: "My hens are not half-starved. They are well-fed, heavy, and, if anything, are too fat." No doubt they are, but remember to starve or half feed a hen does not necessarily imply a lack of quantity of

55—A GOOD ARRANGEMENT TO ADMIT SUNLIGHT

grain supplied her. One hen may be reduced to skin and bone, starved for want of food—another may be rolling in fat and have stored all the surplus she can carry and still be starving for the flesh, bone and feather formers, and is suffering just as much from starvation as the thin one.

FOR MOLTING FOWLS.

The basis of a good forcing food for chicks is eight bags of oats and one of barley, ground together, with the hulls ground to a meal, to which can be added one bag each of fine bran, corn meal and clover meal, with ten per cent. of beef scraps and ten pounds of salt. This figures a ration of 1:4.5, and mixed with skim milk will put meat on chicks as fast as they are capable of digesting the food, providing not only for the normal growth of tissue, frame and feathers, but furnishing a surplus which the tissues appropriate, resulting in a plumpness of breast which is very desirable.

The above will be found a very satisfactory mash to feed birds in the molt, either morning or night, as preferred, and the alternate feed of whole grain should be as narrow as 1:4. The beef scraps and salt are very important during the molt, although equally important at all times and should never be omitted. Many a fancier actually starves his birds for these elements until they become desperate and denude each other of feathers to satisfy their craving. Should the reader ever find a pen of birds in this condition he can correct it almost immediately by a liberal feed of bologna sausage, which satisfies their craving for salt and meat. Thousands of poultrymen are not aware of this and many other simple facts, and their birds are suffering in consequence far more than they realize. The craving in itself is suffering and just as intense, whether for these elements or for water; one is just as necessary to their health and well-being as the other. Supply material with which they can grow feathers and the process will be an easy one quickly completed without any set-back or disturbance of their health, and many will be found laying during this period. A carpenter can build a house if supplied with all the necessary material, but with all stone and no lumber, or all lumber and no nails, he can do nothing.

WASTE PRODUCTS INTO EGGS

SCARCITY OF GRAIN AND MONEY FORCES US TO UTILIZE EVERYTHING THAT HAS A FOOD VALUE—IN TIME OF PLENTY WE CERTAINLY SHOULD DO THE SAME THING

MRS. S. B. TITTERINGTON

ONE beneficent result of the great drought last summer was that it compelled the poultry raiser to supplement the scarce and costly grain foods with other material. Never, perhaps, have the so-called waste products of farm and household been subjected to such close scrutiny in regard to their availability as in this emergency. The knowledge thus obtained under trying conditions will be turned to profit in coming years.

In the list of waste products we must place first and foremost green bone. While the use of this valuable adjunct to poultry rations has been known and urged for years, many were slow to give up the convenient and easy grain method. The balanced ration was a term that had little meaning to many busy people in other days; but it is safe to assert that the value of what the term stands for is better understood to-day than ever before.

So much has been said regarding the value of cut green bone that it will be impossible to advance any new ideas along this line. The assertion that green bone is an egg in a different form, or, in other words, that green bone contains the necessary constituents for eggs, feathers, flesh and bone, is supported by results. As a supplementary food for winter egg production, enhancing fertility, as well as the number of eggs, it is easily in the lead. Bone cutters are legion, and among the multitude offered, there surely should be satisfactory machines. But for a few hens, where no bone cutter is available, a chopping block and a hatchet make possible this highly esteemed addition to the food.

For growing chicks green bone is invaluable. It insures

56—INEXPENSIVE PORTABLE HOUSES

quick growth and sturdy fowls. Of course an over supply will work disaster, but used with care and judgment it will do wonders.

By green bone is meant a strictly fresh, untainted article. Boiled bones, or bones that have lain out in the sun and rain, will not answer. Avoid also the bones of animals that have died from disease or starvation.

FEEDING FOR EGGS

Table scraps are especially good for poultry. They may form, along with potato or other vegetable peelings, the basis for a warm mash. By carefully saving everything of this description, so that enough of the waste material is cooked together to make the bulk of the mash, it will not require as much ground food for making it the required consistency. It should never be sloppy, as this is prejudicial to health.

GREEN FOOD

Green food can also be supplied from the store of cow beets or mangel wurzels, usually grown for the cattle. Cabbage leaves are a highly esteemed relish. Almost anything in the vegetable line may be utilized for the hens. Small potatoes, turnips and carrots are good things to have in days of scarcity. The cow beets are not exactly waste material, yet the small and imperfect ones are just as acceptable in the poultry house as the larger ones which the cattle can dispose of more readily.

No egg shell should ever be burned. They are too valuable to dispose of in this way. Crush them and add to the mash. They will help to supply the line for future eggs.

The by-products of the dairy are too important to be overlooked. Skim milk, butter milk and sour milk are all nourishing and stimulate egg production. It will be money in pocket to make the pigs divide with the poultry. It will be more profitable in the end.

Butchering time affords a grand feast from otherwise waste products for the fowls. Hog livers, kidneys, and the like, are soon disposed of if put where the poultry can pick at them. Beef heads seem to be particularly popular in chickendom. After every particle of flesh has been cleaned from the outside split them open, if you have not a bone cutter. The inner part contains much which the busy foragers can make good use of. If you are the happy possessor of a bone cutter, do not throw aside the leg bones because hard to cut. They are especially rich in the constituents for which we feed cut bone. Have coal ashes within reach The fowls pick from them many tiny particles which aid digestion. Boxes of road dust are a necessity in winter. The hens delight in their dust bath when confined even more than in their summer liberty. By adding a little good lice powder to the dust, you may keep the fowls free from insect pests.

BRAN, SHORTS AND LINSEED MEAL

Bran, shorts, linseed meal—the by-products of flour and starch—may be bought at mill or factory at a price much below their real food value. They are, of course, far cheaper than the whole grain, and may be made to largely take its place by mixing in mashes, etc.

GRIT

Grit is surely a waste product, as it is good for nothing save as a grinder in the digestive process of our feathered pets. Many people pound up broken dishes, glass and the like for grit. The writer can not speak from personal experience of its utility, but strong claims are made for its value. But grit, whether home-made or commercial, is an imperative necessity if we would have our poultry in the best of condition—that health which insures a profit as a return for our labor.

Enough has been said regarding waste products to suggest the wisdom of looking out carefully for such as may prove adapted to the needs of the poultry. Doubtless there are others, which some investigating person will discover from time to time. In all our work, let us heed a scriptural injunction, "Gather up the fragments, that nothing be lost."

TABLE SCRAPS INTO EGGS

ONE HUNDRED HENS IN CONFINEMENT FROM NOVEMBER TO JULY TURN IN A NET PROFIT OF $370.82—A UNIQUE BUILDING AND ORIGINAL METHOD OF CARE AND FEEDING

CHARLES STEWART

IN CITIES and large towns where natural gas is used exclusively as fuel the refuse to be carted away as garbage differs materially from that of cities where different kinds of fuel are used. In the latter, ashes form the bulk of the waste matter, while in the gas-consuming towns where the refuse cannot be burned in the stoves, table scraps, etc., form the main portion. In the suburbs of such cities the garbage collectors sort out these table scraps for use in feeding hogs. A few chickens are fed, but it is a field almost uncultivated in the interest of poultry.

For the benefit of your readers, I will give the result of the experience of a "feeder," who is now keeping 100 hens for eggs. As this is his third year in the experiment, the facts can be tabulated and known results recorded. He has been feeding hogs on table scraps for several years, but two years ago last fall decided to try a few hens.

Aside from the two by four hemlock scantling and some glazed sash that was bought, the material of the house is box-

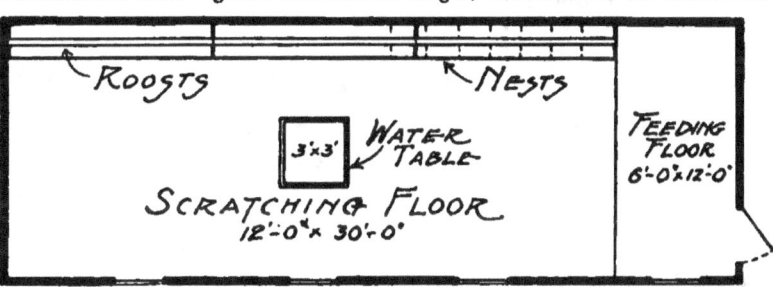

56—PLAN OF HOUSE FOR UTILIZING TABLE SCRAPS

boards. It is a shed-roof house, 7 feet high fronting south 5 feet high in rear, 12 feet wide by 36 feet long. Along the north wall, 3 feet from the floor, is a droppings board, 18 inches wide and 6 inches above it is a roost. For nests, orange boxes are placed on the floor under the droppings board, set out a little from the north wall. Slats are nailed at the front and top, with space for the hens to enter at the back of the box. The only door in the house is in the east end. Six feet of the floor at the end of the building nearest the door is partitioned off by a board about 14 inches high and is used as a feeding floor. The balance of the floor is covered about a foot deep with buckwheat hulls—a waste product costing nothing. Near the center of the house is a table 3 feet square, 24 inches high, on which stands the water-pan. There is no dust box, grit box, feed trough or other furniture, The hens use the hulls as a dust bath. The upper sash of one or more windows is opened every day for ventilation as the hens are never let outside the house.

SELECTING THE HENS

The hens are bought in October from live poultry displayed at the markets, from farmers' wagons and from hucksters. They are not bought in bunches, but are picked out seperately on their individual merits. A hen showing debility is rejected, while a white ear-lobe, as the mark of the non-setter, is a favorable point. Well-grown pullets are preferred, but vigorous hens that have gone through the molt are considered almost equally as good. Where a hen has a full coat of feathers and there is a doubt as to whether it is an

EGGS AND EGG FARMS

old or a new growth, it is decided by plucking a feather. If new, it will show a little blood at the point of the quill.

CARE AND FEEDING

As the hens are secured they are treated for parasites, placed in the house and are fed on cracked corn with plenty of fresh water. Then the house is stocked to its capacity—one hundred hens—the feeding of table scraps (supplemented by cracked corn) is commenced, and continues without change until the hens are sold the following summer, about July 1st. As no mash is fed in which to give condiments, the hens are "medicined" after the Scotch fashion—or Harp method—in the drinking water. Tincture of cantharides is given every day for one week, the next week common black gunpowder is given and the third week the tincture of cantharides. By the end of the third week they will practically all be laying and after that the cantharides and gunpowder are given but once a week alternately. Cracked corn is fed every day. In the morning four quarts of the corn are scattered in the hulls on the floor and at noon a liberal supply of table scraps (two, three or four bushels) are thrown on the feeding floor. In the morning they are busy in the buckwheat hulls, digging for the corn; most of the afternoon will be spent picking at the scraps on the feeding floor, where they find meat, potato and apple parings, celery tops, cabbage, tea leaves, bread, etc. At dusk the attendant gathers up whatever bones there may be into a wire basket, and sweeps up the balance of the scraps to give to the hogs, leaving the feeding floor clean for noon next day.

BURNED BONES FOR GRIT

The wire basket containing the bones is hung over a gas fire and when they are thoroughly burnt they are put through a grit mortar and thrown into the scratching litter in lieu of grit, lime and charcoal. On this treatment the fowls keep remarkably healthy, not one having died since the first winter.

PROFITS ON HENS AND EGGS

As the hens are bought in October and sold in July, at the end of the experiment they weigh a little more and bring a little higher price per pound on the market than in October. The eggs are marketed without any expense, two or three times a week. Last year, during the eight months from November 1st to July 1st, the amount received from the sale of eggs was $388.32, or an average per month of $48.54. The prices ranged from 22 cents to 46 cents per dozen, or an average of about 30 cts. per dozen throughout the eight months. The only food bought was seventeen hundred pounds of cracked corn, average price delivered, $1 per cwt. The "medicine" cost less than fifty cents. This leaves a net profit of $370.82 on the 100 hens for the eight months, or an average of about $3.71 per hen

HEAVY LAYING AND FERTILITY

ARGUMENTS AND EVIDENCES INTENDED TO PROVE THAT THE TWO HUNDRED EGG HEN IS FULLY AS REPRODUCTIVE AS HER SISTER, THE POOR LAYER—CORRECT AND ECONOMICAL FEEDING AND BALANCED RATIONS—STUDY NATURE'S METHODS

A. J. SILBERSTEIN

THOSE who have kept complete records for three or more years are prepared at any time to furnish sufficient facts to prove that the heavy layer is by no rule identified with the infertile hen; that, for some (as yet) unexplained causes, all flocks have their fertile and infertile hens—hens whose eggs hatch well and hens whose eggs are almost entirely infertile; that in both the fertile and infertile classes there are heavy, and average, and poor layers; but to assume that heavy layers' eggs are infertile because of heavy laying, is as wide from the facts as the statement that poor layers' eggs are infertile because of the few they lay.

The causes that induce fertility and infertility are almost entirely unknown; and if I were to hazard a guess from my own efforts to obtain light on the subject, I should be prompted to state that it will be a long time before information on this important question will be had, unless a lucky accident discloses the truth, as a fall, or idle kick at a stone has in the past disclosed a rich mine.

Where stimulation is resorted to for increased egg yield, it is doubtless to be expected that the eggs from the heaviest layer would prove the most infertile. There are many hens which cannot be stimulated to increased egg yield, and when condiments are regularly fed to such, infertility is immediately noticeable in decreased percentage of eggs hatched and increased percentage of mortality in chicks. Other hens, again, respond readily to stimulation, and at an even greater cost in fertility; the heaviest layers under such conditions being naturally the poorest in fertility. This is no guess, but a lesson learned from experience, and one that any breeder can readily verify at little cost in time.

Without knowing more of the experience in the case of the experiment of Orono station than appeared in print, I venture the assertion that if Professor Gowell were asked as to relation of percentage of fertility to laying (providing fertility records were kept of all layers, and stimulation were not resorted to) he would say that the heavy layer with the large percentage of fertility, was not exceeded in that percentage by his average or poor layers, or both; and that there were as many or more average and poor layers with low fertility records in his flock, as there were heavy layers with that failing. Where forcing is not resorted to, there is no line of fertility drawn between the heavy layers and others unless prolific; nor does heavy (but natural) laying in any way interfere with the productivity of the layer.

We find hens arrayed in classes without regard to their laying ability; some that will strongly stamp their progeny with their individuality; others that seem to reproduce themselves only at times and still others that completely fail in reproduction; and in each class, as in the fertility classes, the heavy, the average, and the poor layers are to be found, proving that natural laying has no influence whatever on reproduction.

TWO-HUNDRED-EGG HEN MORE COMMON THAN SUPPOSED

In some poultry publications I have read articles that seek to prove that the heavy layer is undesirable for breeding purposes, the argument being advanced that nothing is heard in the case of individuals or flocks that have made "phenomenal" (?) records, of their own progeny having equalled or improved the "original" record. I fully realize that a wide field for misrepresentation, if not fraud, in statements of

FEEDING FOR EGGS

prolific laying is offered those with an elastic conscience, and also understand the kind of reception statements of seemingly large records receive from readers of poultry literature. Taking the subject as a whole, I assume responsibility for the statement that, were individual records of laying more generally kept, the two-hundred-egg hen would be found to be quite numerous—perhaps seven to ten per cent. of the pure-bred flocks would be found to exceed that mark; but large flocks that exceed an average of one hundred and fifty eggs per hen per year are extremely rare, if such exist.

It has been argued that if the two-hundred-egg hen can reproduce herself, it should be an easy matter to obtain a pen of her daughters, all two-hundred-egg hens, the first year; that this pen could easily produce a hundred two-hundred-egg hens another year; and so on, until the absent-minded theorist has succeeded, in fancy, in making eggs worth about ten cents a carload. On this same basis a ninety-five-point hen ought to produce a pen of ninety-five-point pullets the first year; this pen of pullets a hundred ninety-five-pointers the second year, and so on, but they do not, and probably will not until we who breed them learn some of the A B C's of breeding, and I fear that day will not dawn in the lives of the present generation.

That the two-hundred-egg hen does reproduce herself quite as often as, and oftener than, the ninety-five-point hen, is well known by those who have kept records for a few years. Just as breeders find it an easy matter to reproduce the ninety-two-point breeder—harder to reproduce the ninety-three-point bird—difficult to reproduce the one that scores ninety-four, and a rare occurence to reproduce the male or female that reaches ninety-five, just so is the one hundred and twenty egg per year hen readily reproduced, the one hundred and fifty egg hen not quite so readily, the one hundred and seventy-five egg hen quite difficult, and the two-hundred-egg hen about as easy to get a flock of as it is to get a flock of ninety-four-point hens. That "like begets like"—with intelligent allowance for reversion—is the experience of all old breeders in breeding all classes of pure-bred stock.

To those who keep records, the above statement need not be proved, for each of them have been overwhelmingly proved in his own experience. To convince others, I give a brief statement, copied from my records, showing the effect of heavy (?) laying on fertility and the reproductive powers of so-called heavy layers. Unfortunately the records cannot be given for the year because of circumstances beyond my power to control. Keeping records was rendered impossible from September 12 to October 14, and parts of former years' records were lost. I give the statement without further comment:

Dam's Record Season of 1899-1900	Pullets Retained for Breeding, 1900	Pullets began Laying	Record to Sept. 12	Total Eggs Set	Total Chicks Hatched	Sold	Alive Sept. 3
233 Eggs...	T 23	Feb. 10	a134	46	22	10	4
	T 24	March 17	b 83	31	8	0	5
	T 89	Feb. 8	148	31	27	15	7
	T 105	March 26	146	10	6	1	5
	T 161	Feb. 19	173	30	24	8	11
	T 180	March 20	c111	35	0	0	0
209 Eggs...	T 15	Jan. 26	157	33	26	6	14
	T 150	Jan. 20	181	40	7	0	2
205 Eggs...	T 51	Feb. 5	b139	32	5	0	0
	T 77	Jan. 9	194	52	45	17	22
	T 85	March 5	151	27	15	4	3
197 Eggs...	T 84	Dec. 31	189	52	38	20	9
	T 70	Feb. 17	142	23	6	1	1
	T 149	Feb. 10	d 67	7	3	0	3
195 Eggs...	T 100	March 24	125	22	15	5	6
	T 192	Jan. 6	194	33	19	0	13
231 Eggs...	Eggs infertile during all of first year						

a T 23 died July 28. *b* T 24 and T 51 sold Sept. 4. *c* T 180 sold August 1. *d* T 149 sold May 5.

That feeding has an important bearing on egg yield is too well known to require argument; but not in the sense in which it is very often discussed. Correct feeding is an essential to perfect health, and perfect health is essential to success in breeding. To force a fowl, and continue the forcing beyond her natural ability, will meet with the same results that an effort to continue forcing a race Horse to do the work of a draught horse would. With hens it does not always result in the death of a fowl, although I am quite sure that the breeder is fortunate who so quickly gets his proof. The results of forcing are too often evident in reduced vitality of the second and following generations, and that is the most expensive of all experiences in this line. With correct methods in feeding we can only hope for improvement in laying, and in shape, size, feather, etc., by rigid and careful selection.

What is correct feeding? Who among us knows? Speaking for myself, I am most anxious to learn. My construction of the term "correct feeding" implies the method which shall at lowest cost of material and time, keep fowls in perfect health, as is evidenced by an ample yield of hatchable eggs, while maintaining the bird's size (weight) and uniform weight

47—WHITE PLYMOUTH ROCKS

of eggs. We frequently hear of instances where fowls of no particular breed, and receiving no care but an occasional feed of grain (generally corn), are yet giving good returns. If this is "correct feeding," we want to know it without delay. I had the opportunity of investigating one such case that was brought to my notice. It can not, of course, be taken as a criterion for all, but such as it is I give it, believing that it sheds some light on the subject.

In the case referred to the fowls had free range, access to manure cellar and every other place their fancy prompted; roosted whever they pleased, were fed leavings from the table and when these were scarce, were given corn. About forty fowls were giving eight to twelve eggs a day at the end of November. It was my good fortune to call when eggs were being gathered, and I counted eleven. I asked how often eggs were collected, and was told: "Every now and then." Further questioning brought out the fact that no one had gathered eggs Thursday (Friday when I called) or Wednesday, so here we have the egg yield dwindled down to one-third.

KEEP CLOSE TO NATURE'S PLANS

In my effort to obtain results which seemed to me satisfactory, I have endeavored to imitate nature as far as my limited

EGGS AND EGG FARMS

knowledge gave me a conception of her methods, and to improve on them where, in my judgment, improvement was possible.

In observing wild birds it has seemed to me that they hunt food about all day long, and in my fancy I have pictured them as often going to their roosts at night with their hunger but partly appeased.

Fowls will not exercise for the sake of exercise; given a full crop and they will doze until hunger prompts them to move. This has been my experience and the experience of all with whom I have conversed on the subject. Given a mash in the morning they apparently do not see the grain thrown in litter for them to scratch for.

With my first lot of fowls I followed the beaten paths given in poultry literature of the day, and fed "all they would clean up quickly" of mash in the morning as instructed. "All they would clean up quickly" bothered me a bit, for I found their appetites to vary considerably, a pen of twelve fowls

58—FEEDING TIME IN THE POULTRY YARD

cleaning up anywhere from one to six pounds, so that, when making the mash, I was at a loss to know how much meal to use. The noon meal of grain scattered in litter seldom interested them, and their scratching was spasmodic and rare. I changed the bill of fare and fed them mash for the noon meal, and after a while secured exercise in the forenoon. Another change, and mash was fed at night, and by this method was secured constant exercise throughout the day. This seems to me more in line with and perhaps an improvement on nature's way. A careful sprinkling of small quantities of grain in deep litter during the day imitates nature in that it compels the fowls to seek for their food grain by grain, while the feeding of mash at night is the improvement on nature's way, which insures a full crop daily just before going to roost. I prefer mash to grain at night, because it digests quicker, bringing birds from the roosts the next morning with a sharpened appetite, while a full grain feed is often but partly digested in the morning. One night an accidental dropping of grain, after they had eaten all they would of mash, surprised me by causing the fowls to jump for it greedily. I thought it over and the next night fed about half of the quantity they cleaned up the night before, then gave another portion, and a third, with about ten minutes' interval between each, and noted they ate fully one-half more in this way. Since then I have repeatedly tried feeding them at once the full quantity they ate the night before, but I never had them finish it. In short, by feeding them in small quantities at a time, I was coaxing them to eat more than they otherwise would. I have never been able to see anything but improvement in results, as a consequence of this method of feeding mash at night, and it has always obtained here since.

THE PROBLEM OF FEED

Next, with me, came the problem—what to feed. In seeking light I dived into the intricacies of "chemical analysis" and "nutritive ratios." Taking as a basis for my figures an average of analysis of each of the grains I used, I found my bill of fare approximated about 1:9—one part protein to about nine parts of the fats. Upon the assumption that feeding laying fowls might for the purpose in view be compared with feeding milch cows that best results from the latter were had when fed a ration approximately equaling in nutritive ratio that of milk, their product, I assumed that best results in laying might be had from fowls if fed a ration equaling in nutritive ratio that of the egg, their product, and in line with this theory I attempted to compose a bill of fare, the chemical analysis of which would show a nutritive ration of about $1:2\frac{1}{4}$.

I have at different times bought mongrel hens that obtained the greater part of their food on free range in the spring and on opening their crops found them filled mostly with insects and worms, quite a little green food, with seeds, grains, etc., in small quantities. In the absence of a knowledge of chemistry I have assumed that this fact confirmed my theory of a narrow ration ($1:2\frac{1}{4}$ or lower) for laying hens, believing that analysis would show the animal food found in crops to have a value akin to lean meat.

In attempting to feed a ration approximating in its entirety a value of $1:2\frac{1}{4}$ I found the need of a very narrow mash to overcome the high nutritive in the various grains. In narrowing the mash I at first added considerably to its cost, because of the quantities of expensive concentrated foods needed. Again, if a narrow ration is correct, it presents an added reason for feeding the mash at night, because of the quantity that needs to be fed, to overcome the total of fats in the grains given the fowls—a quantity which if fed at any other time would effectually stop exercise for the day. The trouble encountered at first was in the expense, and in loosening the bowels to a disagreeable extent.

As before stated, to obtain the nutritive ratio of the rations I fed, I used the figures given as the average of the grains in the various experiment station reports. The fact is that each of the grains vary widely in their chemical analysis, and I have no practical means of ascertaining how near each kind of grain I buy approaches this average, nor, for that matter, how near or wide apart each lot is compared with the last. For the reason, I have of late paid no attention to the actual figures bearing on nutritive value, except to accept the general fact that, for instance, bran figures about $1:4\frac{1}{2}$ wheat 1:8, etc., framing any new bill of fare that

I may compose on that basis, aiming to keep the whole ration approximately near to 1:3, which figures (based on averages) were the last I took the trouble to ascertain, and which has given me better results—fewer fat fowls and better digestion than any other, but it still leaves much to be desired.

WINTER EGG PRODUCTION

A VERY INSTRUCTIVE ADDRESS OF A LEADING POULTRY MANAGER — PRODUCING A QUICK MOLT — ANSWERS TO OFT-REPEATED QUESTIONS — DO NOT FEAR COLD AIR

W. R. GRAHAM

THE main essentials to winter egg production are as follows: Good stock, comfortable quarters, proper food, reasonable exercise, cleanliness and favorable weather. As to the stock for winter egg production, April and May hatched pullets will give the best results. If they have been well fed during the summer they are almost sure to lay in December, January and February, if properly managed. As to breed, I regard Wyandottes, Rocks, Orpingtons and the American breeds as the best. The question of early molting must be considered. Whether it pays to have hens molt in July or August depends on the local conditions, on the price of eggs in the fall, and on the cost of production in summer as compared with winter. The molting season may be controlled in this way: Give the hens you desire to molt one third of the amount of food they have been accustomed to receive and you will stop egg-production. After that treatment has continued about two weeks, feed them well, and they will either lay better or they will molt. Seventy-five per cent. will molt under this treatment.

THE INDIVIDUALITY OF THE HEN

Most people do not appreciate the difference in the individuality of their hens. The hen that is thrifty, vigorous and full of vim—the business hen—is the kind you should select. A great deal depends upon the strain. I can give you strains of Plymouth Rocks which I can almost guarantee not to sit; other strains I can guarantee to sit frequently. If you have a family of admirable egg producers, place a male of that strain in your flock, and the egg production will be increased in his progeny. Ability to produce eggs or to produce flesh can be bred in poultry just as surely as the ability to produce milk or beef can be bred in live stock by the careful breeding and selection.

THE GRAINS FED

It is said that "feed is half the breed," but in my opinion the breed is the big half. Still it is essential to know what to feed hens for egg-production. Wheat is the best egg-producing food among the grains, but it is doubtful whether it can be fed with profit when it is very high in price. In the ration we should have grain, vegetable food, meat food and water. In grains we have found a ration of equal parts of wheat, barley and oats to be very satisfactory, but the oats should be a thin hulled variety. Corn is a good food in cold weather for Leghorns, but it is not food for breeds of chickens that tend to put on flesh early. Vegetable foods I would place in the following order: Mangels, cabbage, turnips, sugar beets. The best way to feed roots is to stick a mangel on a nail in the wall, as they will then eat practically the whole of it.

We feed about two lbs. of whole grain at 9 a. m., burying it in six to twelve inches of straw litter, placing the grain first and the straw on top, so as to give the birds exercise. At the same time we open wide all the windows so as to admit the fresh air. We follow this with a feed of either barley or meat, and we always give roots. We usually feed meat three times per week. Some times we feed them bread and milk by way of variety. We keep beef scraps constantly in front of them. This is a by-product of the abattoir. In addition to this we give them one pound of meat three times a week. Beef scrap is one of the strongest nitrogenous foods we have. At night we give them whole grain or a mash.

Q. Do you feed clover?

Mr. Graham: I feed steeped clover leaves in a mash.

Q. Do you feed the mash warm?

A. Just warm. I do not think it is essential to feed a mash, but sometimes it is a convenient form in which to present a ration. The meat food may consist of waste from the kitchen or slaughter house, or blood meal or animal meal may be used depending on circumstances. Beef heads at 20 cents each form a reasonably cheap meat food. They should

59—A CONVENIENT STYLE OF NEST BOX

however be boiled or ground in the bone mill, and if for grinding should first be frozen hard and the teeth extracted.

Q. Will milk take the place of meat?

A. It will in summer, but it does not give good results in winter. It contains so much water as to lower the heat of the body.

THE HOUSE MUST BE SANITARY

If you wish to secure winter eggs, you must pay attention to keeping the poultry house clean. If you cannot clean the droppings board once a day, clean them once a week and throw some ashes on them. White-wash either with a broom or a spray-pump, and choose a dry time for the operation, or it will take a long time to dry, and the dampness is apt to give the chickens roup. To kill lice on the hens, sulphur and insect powder is a good remedy. Go over the pens once in every two or three weeks with a mixture of coal oil and carbolic acid, in the proportion of one pint of acid to a gallon of oil.

A reasonable amount of exercise is necessary to keep the hens in good condition. Two hours per day, picking over the straw, will be found to be sufficient. There is a difference between exercise and work; what is exercise for the Leghorn

EGGS AND EGG FARMS

will be work for the Brahma. Do not use up the food you give them by making them work too hard, but on the other hand, do not allow them to become too fat.

Q. In feeding liver, lights, etc., do you cook these foods?

A. Yes, for the reason that there is less danger of disease when cooked. It is necessary to give hens egg-shell material in winter time, and there is nothing better than egg shells when they can be obtained, as they usually can be from a hotel. This is one of the best cures of egg-eating.

Q. In most poultry houses on the farm, the roosts are built step fashion. This is the worst kind of a roost for the reason that it is a natural characteristic of the hen to climb to the top, and in such cases the top roosts are always overcrowded. This results in overheating and chills follow. The roosts should therefore always be on the same level. Where there is a drop curtain in front of the roost, it is my practice to have it extend right to the roof. Do you think that is desirable, Mr. Graham?

A. The question of curtain or no curtain is a debatable one, I incline to the opinion that a curtain is apt to make the roost too warm at night. (When the curtain is lowered only on extremely cold nights there would be no danger of the quarters becoming too warm.—Editor.)

POULTRY MUST BE GRADUALLY ACCUSTOMED TO COLD AIR

I believe in fresh air for poultry, but at the same time the birds must be brought up to it and accustomed to it gradually as the weather gets colder in the fall. Sometimes the farmers, being impressed with the importance of fresh air for their poultry through listening to an address at the Farmer's Institute meeting or elsewhere, suddenly revolutionize the treatment of their poultry in this regard, and the birds take cold as a result.

In the average flock of poultry on the farm, there are far too many varieties. You cannot secure good results in winter egg production under these conditions. The characteristics of your fowls should be uniform as possible, no matter for what purpose they are kept. Treatment and food that will place one breed in the proper condition for laying, may have quite a different effect on another breed, and it is therefore important that one breed only should be kept.

EGG-EATING CURED

TWO TRIED REMEDIES FOR CURING A PERNICIOUS HABIT—METHODS ARE DIFFERENT, YET EFFECTIVE

MRS. ROBERT WALDRON

I have found an easy remedy for the egg-eating habit among hens. Every year in February my Brown Leghorns have seemed to get into vicious ways, and having once begun to eat eggs in a few days I would scarcely be able to get an egg from the Leghorn pens, unless I happened to be right on the spot when it was laid. For some years I could find no way to prevent it, till time brought relief and fine spring weather enabled me to get them out on the ground. For the last three years, as soon as the egg-eating contest begins, I go to a bakery and get a large lot of egg shells. Sometimes I get a barrel full at a time. For the last two weeks I have been getting them nearly every day, and never less than half a

60—A FRESH-AIR POULTRY HOUSE

sugar barrel full at a time. They are thrown by the bushel into the pens, and egg-eating has entirely ceased. It is the long winter's confinement that causes the trouble, I think, and the birds are not to blame.

MR. CONRAD ORTTENBURGER'S METHOD

This is my experience with egg-eating hens. I have a flock of about forty; well housed and cared for. They are fed a variety of grain, green cut bone, soft food and plenty of grit, etc., also green food. My fowls look well and have been laying well, but about two weeks ago they took a notion to eat eggs. As soon as a hen laid, the balance would pile in and devour the egg; consequently I got no eggs at all. I was at a loss to know what to do, I had heard that the hatchet is the only remedy, but could not make up my mind to use it, just at the laying season. So I though I would try another remedy. I saved up a panful of shells as nearly whole as possible, soaked them in coal oil and sprinkled powdered alum over them; then I put one shell in each nest. The first two days they ate them; the third only partly, the fourth not at all. I then discontinued putting shells in the nests. It is nearly a week since I did it and I get plenty of eggs and no sign of egg eating. I am positive they are cured of the habit.

EXPERIMENTS—LAYERS AND EGGS

EXPERIMENT ON FEEDING FOR EGGS

FOUR LEGHORN PULLETS LAY AN AVERAGE OF ONE HUNDRED AND EIGHTY-TWO EGGS EACH PER YEAR AND SHOW A PROFIT OF TWO HUNDRED PER CENT—FEEDING BASED ON THE COMPOSITION OF THE EGG—INVESTIGATION IN REGARD TO FOODS

JAMES DRYDEN

ONE thing that has been demonstrated by the Utah Experiment Station is that there is money in hens. A profit of 200 per cent was made on the cost of food during the year. That is, a pen of four Leghorn pullets laid an average of 182 eggs each during the year, which began and ended in November. It cost during that year 62 cents to feed each pullet; that is, the food consumed charged at regular market prices cost 62 cents. Wheat, which was about half the cost of all the food, was charged at 70 cents a bushel. Eggs were sold at market prices in Logan. For several months they were ten cents a dozen and one month they were 25 cents. The 182 eggs at those prices were worth $1.88. That is a profit of $1.26 per fowl. These results were confirmed by the experiments of the subsequent year.

It is safe to say that at the present prices of eggs and food, a profit of from $1 to $1.50 on a food cost of 62 cents is quite within the reach of intelligent management. I say intelligent management. It cannot be done, however, with dunghill fouls. It cannot be done with birds that have been bred for the show ring for several generations; they must be bred for eggs just as the Jersey has been bred for milk. Nor can it be done with hens that have discarded their teeth. The hen after carrying on an egg laying business for two years has done her part to civilize the world and after that becomes a back number. Neither can the 200 per cent performance be reached unless the chick is hatched early enough in the spring, so that by the end of October or November she is ready for business.

What to feed is probably the most difficult of the many problems in successful poultry culture. In studying the question of feeding we ought to begin by studying the egg, just as a manufacturer of a good plow knows exactly what materials are necessary for the completed plow before he begins work on the raw material. The hen is an egg factory, and we must put the right kind of raw material into the factory if we expect to get eggs from it. We should study the composition of the egg which is composed of several constituents. A great many people forget that. From the way they feed the hen they seem to think it contains only one chemical compound. Many a man feeds the hens one variety of food, and if he does not get an egg by feeding on such a ration he declares there is no money in hens. Then he goes back to sheep farming and hog raising, cow farming and politics, and of course that is a good way back.

Let us consider the composition of the egg, and then draw some conclusions as to how to feed for eggs. An average egg will weigh two ounces; of that, 10.81 per cent is shell, 32.47 per cent is yolk, and 56.42 is white, according to a California analysis. The shell is nearly all lime, or carbonate of lime. The yolk is composed of 50 per cent water, 15.5 per cent protein and 33.4 per cent fat, and about 1 per cent mineral matter. Of the white, 86.48 per cent is water—a very cheap commodity—12.07 per cent protein, .23 per cent fat, and .34 per cent mineral matter, of the total weight of the egg 65 per cent is water. More than an ounce of water, therefore, is stored away in every egg.

Of course that does not mean that she must drink that amount of water, for a certain percentage of all poultry food is water. For instance, in wheat there is between 10 and 11 per cent water; in other words, a hundred pounds of wheat contains 10 to 11 pounds of water. So, if a bran mash is fed, about half of that is water. By a little figuring we discover that the egg contains a quarter of an ounce of protein. How is the hen going to obtain that protein, supposing she is fed as she very often is fed, on wheat and water?

WHEAT AS A FOOD

Now there is in wheat about 12 per cent of protein, or 12 ounces of protein in a hundred ounces of wheat. Supposing the hen weighs four pounds and that she eats four ounces of wheat a day—which she probably would not do very long. Supposing, further, that it takes three of those ounces of wheat to maintain the the body, there is one ounce left, which the hen is very willing to convert into an egg. In one ounce of wheat there is .12 of an ounce of protein, and we have seen that the egg contains .24 ounce or practically one-quarter of an ounce of protein. So that there is just half enough protein in one ounce of wheat to make an egg. If the hen, however, were inclined to lay an egg only every other day, she could make half the egg with it and wait till the next day to get the requisite amount to furnish protein for the other half. An egg every other day is not so bad.

But the egg contains something else besides protein. It contains .23 of an ounce of fat, or nearly as much fat as protein, not quite. Now that same ounce of wheat from which we just got half enough protein to make the egg, contains .72 of an ounce of carbohydrates and .02 of an ounce of fat, or say three-quarters of an ounce of carbohydrates and fat, or just three times as much as the egg contains of fat. You see what a quandary you put the hen into by feeding a straight wheat ration. You give her just half enough protein to make the egg and three times as much carbohydrates and fat as are necessary. What is the hen going to do about it? She is an honest hen. Honesty has been so bred into her through all the centuries that she could not make an egg as some of the human bipeds make the filled cheese. She could not, if she would, take some of the surplus carbohydrates and fill them into the places where the protein ought to be. If you could by some means induce her to eat an ounce more of wheat she would then have

61—FOUR FAMOUS LAYERS

the requisite amount of protein, but she would then have six times as much carbohydrates and fat as are necessary. In that case five-sixths of the carbohydrates must be either wasted or the protein diverted into other channels.

But what about the shell? In the ounce of wheat there is only half as much shell material as is necessarry for the egg. Like protein, there is a deficiency of just one-half. So that, in this ration of wheat alone, there is in technical feeding language, a lack of balance. There is too much of one element and two little of another. There is too much of the fat-producing elements and too little of the protein and lime. What the hen does with the food in such a case is pretty much of a problem. But it can safely be said, and there is practical experience to bear out the statement, that she will refuse to make eggs. Wheat has what is called a wide nutritive ratio, one of about one of protein to 9.5 of carbohydrates, whereas profitable egg production calls for a ration of about 1 of protein to 4 of carbohydrates.

The only thing for the feeder to do is to furnish some other kinds of food that are known to be rich in protein. Bran is such a food. It has a ratio of about 1 to 3.8, but enough of it cannot be fed with wheat or corn to balance the ration properly. It is all right to feed a little of it, for the hen likes a variety. Some other food must be added which is still richer in protein, such as lean meat. Cottonseed meal is also rich in protein, so is dried blood which is nearly all protein. But the question of poultry foods and their composition is too large a question to enter into a full discussion of here. Professor Jaffa, of the University of California, recommends the following standard rations for fowls: For 1,000 pounds of laying hens, of about three to four pounds average weight, the food requirement per day would be from 65 to 70 pounds of total food, or about 52 pounds of water free food, containing 9 pounds of digestible protein, or flesh formers, four pounds of fat and about 20 pounds of carbohydrates or starchy material. Per hen the amount would be 3¼ ounces of total food, 2¼ ounces of water free food and .43 ounces of flesh-formers, and about 1.2 ounces of fat and heat producers.

For one thousand pounds live weight of hens whose weight averages about 6 pounds, the food requirements per day would be 40 to 50 pounds of total food, containing 34 pounds of absolutely dry matter, which should comprise 6 pounds of digestible protein, 14 pounds of carbohydrates and 2 pounds of fat. If we calculate this for the individual fowl we would have 4¾ ounces of total food, 3¼ ounces dry matter, with 58 ounces of protein or flesh formers and 1.54 ounces of fat-formers and heat producers.

HOW MUCH DOES A HEN EAT?

All poultry and stock foods vary in the amount of water they contain as well as in the amount of digestible nutrients. These two things should be understood before the poultryman can decide upon the relative economy of different foods, and they should also be understood before an intelligent answer can be given to the question as to how much food a hen eats. In 100 pounds of skim milk, for example, there are only 10 pounds of food and 90 pounds of water. In 100 pounds of mangels there are only 9 pounds of dry matter and the rest is water, and water is not food. In green alfalfa there are about 20 pounds dry matter per hundred and 80 pounds water. In cured alfalfa there are only about 50 pounds water. With the aid of a table of chemical composition of foods, which may be obtained from the state experiment station the poultryman may figure up the amount of dry matter; in other words, the amount of food consumed with the amount of water eliminated.

In experiments made by the writer at the Utah Station, Leghorn hens consumed an average of 2.5 ounces of dry matter, or water free food per day. Wyandottes consumed 3.2 ounces and Plymouth Rocks 3.5 ounces. This is equal to 57 pounds per year for Leghorns, 75 pounds for Wyandottes and 80 pounds for Plymouth Rocks. This did not include all the green food, such as grass, eaten in the yards.

A pen of Plymouth Rocks which laid an average of 176 eggs per fowl during the year, ate 3.7 pounds water-free food per day, the composition of which was as follows:

Ash .23 per cent; protein .65 per cent, carbohydrates 2.72 per cent, fat .29 per cent, the protein ratio being 1:4.9.

AVERAGE AMOUNTS EATEN PER YEAR

A pen of White Wyandottes that averaged 201 eggs each during the year, consumed 3.3 ounces of dry matter per day, the composition of which was, ash .22 per cent, protein, .60; carbohydrates, 2.27: fat, .23 per cent; ratio 1:4.9.

The Plymouth Rocks mentioned ate durin the year an average of 134 pounds total food per fowl, and the Wyandottes, 112 pounds. Figured in dry matter the Plymouth Rocks ate about 86 pounds and the Wyandottes 75 pounds. This means that the food eaten by the Plymouth Rocks contained 48 pounds of water and the Wyandotte ration 37 pounds water This large proportion of water in the food was due to heavy mash feeding, the mash constituting more than half the weight of the weight of the ration, as may be seen by the following table, which gives the weight of each kind of food consumed per fowl:

	Plymouth Rocks	Wyandottes
Mash	372 lbs.	269 lbs.
Wheat	71 lbs.	102 lbs.
Corn	18 lbs.	38 lbs.
Sunflower Seed	11 lbs.	1 lb.
Oats	36 lbs.	33 lbs.
Barley	22 lbs.	4 lbs.
Cut Bones	39 lbs.	45 lbs.
Total	569 lbs.	492 lbs.
Average per fowl	134 lbs.	118 lbs.

The mash was made up as follows:

	Plymouth Rocks	Wyandottes
Skim Milk	186 lbs.	136 lbs.
Bran and Shorts	81 lbs.	80 lbs.
Oats	20 lbs.	
Corn	12 lbs.	22 lbs.
Wheat	31 lbs.	12 lbs.
Gluten	11 lbs.	
Linseed Meal	13 lbs.	12 lbs.
Beef Meal	10 lbs.	
Beans	8 lbs.	4 lbs.
Dried Blood		2 lbs.
Meat Meal		1 lb.
Total	372 lbs.	269 lbs.

From the experiments at the Utah Station, which covered a number of years and a great many pens of different breeds, it may be concluded that the equivalent of about 60 to 65 pounds grain per foul per year for the small breeds, 85 to 90 pounds for the general purpose breeds, and 100 to 110 pounds for the meat breeds, represents about the average consumption of food.

The amount of food eaten will vary with the number of eggs laid, the ability of the hen to digest and assimilate the food and with the temperature conditions surrounding the fowl.

DIGESTIBILITY THE FINAL TEST

I stated above that before we can intelligently judge the relative economy of different foods fed to poultry, we should know something about their digestibility. It is well known that all foods are not equally well digested by animals and that animals do not possess the capacity to fully utilize or assimilate the different food nutrients. By chemically analyzing the food before it is fed and then analyzing the excrement of the animal the investigators have ascertained the percentages of the different nutriments of stock foods that the animal digests.

In 100 pounds of corn meal, for example, there are some 9.4 pounds of protein and of that 6.4 are digestible when fed to cattle. In the same quantity of corn meal there are some 71.34 pounds carbohydrates and 3.67 pounds fat, and of these amounts 66.3 and 3.4 per cent respectively are digestible. With other foods the percentages digested are different.

Dr. Brown of the Bureau of Animal Industry, U. S. Department of Agriculture, recently published some results of a digestion experiment with poultry of a very interesting nature, and if confirmed by further experiments the results will be of far reaching importance to poultry feeders. In brief he found that the nutrients of corn were more largely digested than the nutrients of wheat.

The difference was very great. Of the protein of corn 83.9 per cent was digested and of wheat 77.4. Of the fat of corn 83.3 per cent was digested and of the fat of wheat only 58.8. The same tests showed that fiber is practically indigestible. Poultrymen find it poor economy, therefore, to buy foods containing a large amount of fiber unless at corresponding re-

62—STANDARD-BRED HOUDANS

duction in price. Oats contain a high percentage of fiber in the hulls, and it would seem poor economy to pay for hulls when the hen gets no nutriment from them.

The final test, therefore, of the value of a food is the percentage of digestible nutrients it furnishes. Owing to the difficulty in making digestion experiments with poultry enough data has not yet been accumulated to form an accurate basis for estimating the value of poultry foods from the standpoint of digestibility, but until digestion co-efficients have been determined for poultry we may accept with some degree of safety the standards established for cattle.

EGGS' AT MINIMUM COST

DIGESTIBLE NUTRIENTS WHICH SHOULD BE FED TO LAYING STOCK TO FURNISH CHEMICAL CONSTITUENTS OF THE EGG AND TO MAINTAIN THE HEN IN HEALTH

JAMES R. COVERT

AS COLD weather approaches and the marketability of eggs increases, the problem of how to increase the yield of that toothsome article becomes interesting. The veteran, the amateur, and the good housewife vie with each other in an endeavor to compound a ration which shall produce the maximum yield of eggs at a minimum food cost The public is awakening to a realization of the food value of the egg. More attention is given the subject of feeding, and the agricultural press is devoting more space to articles on poultry. Some of the experiment stations are investigating and throwing light in many hitherto dark corners. Their conclusions in many cases closely coincide with the teachings of experience, and show conclusively that correct feeding is both a science and an art.

EGGS AND EGG FARMS

If to the sum total of the chemical constituents in the eggs produced during a given season, we add the materials required to maintain the hen in health and activity, we have appraximately the amount of digestible nutrients which should be present in her food. As we all know, the digestible nutrients in food articles vary in amount and quality, and some breeds of chickens return a greater profit in eggs for the food consumed than others. This article, however, is confined to the subject of rations which must be prepared with due regard to the purposes for which the chickens are kept. Thus, if we desire to produce flesh we must feed a ration richer in flesh forming ingredients than if we were feeding for eggs, which require nitrogenous materials. Reports of digestion experiments with fowls are seldom met with, presumably because they are not often undertaken. The public should take an interest in the matter and demand of those expert in the determination of feeding problems the solution of this question.

It is assumed that the nutritive ratio for the laying hen and the milch cow should be approximately the same. Their products closely resemble each other, but their relative actual cost makes milk usually much the cheaper food article for man, especially in the larger cities. The German feeding standard for a milch cow calls for 15.4 pounds total nutritive substance in the digestible portion of her food, these nutritive substances to be proportioned as follows: Protein, 2.5 pounds; carbohydrates, 12.5 pounds, and ether extract, or fat, 0.4 pound. This gives a nutritive ratio of 1:5.4. In other words, to every pound of protein there are 5.4 pounds of nitrogenous materials. The nutritive ratio may be determined by multiplying the ether extract by 2.2, adding to this product the carbohydrates, and dividing by the protein. Each pound of fat or ether extract is assumed to have a feeding equivalent of 2.2 pounds carbohydrates. The author has been unable to find the reports of any experiments determining the amounts of these materials necessary for fowls. For want of definite information on several points he is unable to do the subject justice, but with many apologies and a few misgivings, he will attempt to formulate a ration which shall be practicable for the farmer.

It is usual to feed a ration of soft foods in the morning, with a whole grain ration at night. We will suppose to have our choice of the following feed stuffs: Bran, corn meal, ground oats, oil cake, cottonseed meal, beef and blood meal red clover hay, skim milk, with oats, rye, wheat and corn for a whole grain ration. The following table gives the digestive nutrients found in 100 pounds of each of these and a few other articles.

PERCENTAGE DIGESTIBLE MATTER IN AMERICAN FEEDING STUFFS

FEEDING STUFF	Crude Protein Per Cent	Carbohydrates Per Cent	Ether Extract Per Cent
Red Clover Hay	6.5	34.9	1.6
Alfalfa Hay	7.6	37.8	1.3
Cowpea Hay	8.1	37.3	1.7
Potatoes	1.4	16.1	0.0
Corn (average)	7.1	62.7	4.2
Wheat (average)	9.3	55.8	1.8
Rye	8.3	65.5	1.2
Oats	9.1	44.7	4.1
Bran	12.6	44.1	2.9
Middlings	12.2	47.2	2.9
Cottonseed Meal	36.9	18.1	12.3
Linseed Meal	27.2	31.8	2.7
Dried Blood	59.1	0.0	2.3
Meat Scraps	68.4	0.3	13.5
Skim Milk	3.1	4.7	0.8

For convenience we will mix 250 pounds of soft food at a time, selecting as an experimental ration 100 pounds bran, 50 pounds corn meal, 50 pounds ground oats, 25 pounds cottonseed meal, 25 pounds beef and blood meal (assuming the latter to be composed of equal parts of blood and meat scraps.) These quantities, by reference to the foregoing tables, are seen to contain the following amounts of digestible nutrients: Protein, 45.34 pounds; carbohydrates, 101.90 pounds; ether extract of fat, 11.51 pounds. The nutritive ratio we find is 1:2.8, while the German standard for a milch cow is 1:5.4. Therefore, to balance the ratio we must select some material rich in carbohydrates and fat. In selecting clover hay, we secure a high percentage of carbohydrates and at the same time by properly preparing and mixing the clover with the morning mash we are able to furnish what closely approximates green food. Fifty pounds of red clover hay, added to our ration, raises the nutritive ration about 1:3.00. When skim milk is at hand, a very profitable use can be made of it by mixing the soft food with it. A quart of skim milk weighs about two and a half pounds. By adding in the feeding period an aggregate of 100 pounds of milk we make it very palatable but lower the nutritive ratio to 1:2.76. This we will accept for our morning mash, feeding what each fowl will clean up quickly. For our whole grain ration, we may select corn, wheat or rye, as they are all relatively rich in nitrogenous materials and will help balance the ration. We will select corn to scatter in the litter in the evening. If we use two hundred pounds in connection with the two hundred and fifty pounds of soft feed, our nutritive ratio will stand 1:4.3—still somewhat narrower than the standard, but very practicable.

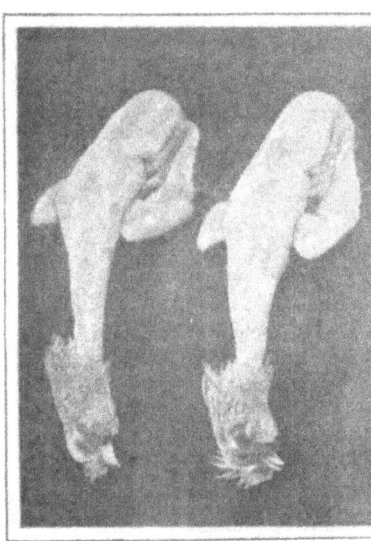

61—FOWLS DRESSED AND SHAPED FOR THE LONDON MARKET

The relative amount of grain and soft food used varies with different individuals, some using more and others less. The nutritive ratio, however, should conform more closly to the standard than the average ration does if best results are desired. The experimental ration outlined above is not intended as a criterion, but simply to show how the different factors are obtained. Theoretically it would be better for the growing chick than the laying hen.

EGG PRODUCTION INFLUENCED BY FOOD

AN EXPERIMENT ON THE FEEDING OF CUT GREEN BONE AND ANIMAL MEAL—RESULTS SHOW ANIMAL MEALS PREFERABLE TO CUT GREEN BONE—VEGETABLE VERSUS ANIMAL FOODS—THE USE OF CONDITION POWDERS FOR LAYING FOWLS

WILLIAM P. BROOKS

IT IS the purpose of this article to bring to the attention of the public the results of several experiments carried out under my direction at the Hatch experiment Station during the past five years, which it is believed are worth the consideration of all who keep fowls for eggs. These experiments have been directed to the elucidation of but a few of the many questions which present themselves to every intelligent person who cares for hens, and this article, which will be based upon the results of our own work, makes no pretense to being a comprehensive paper upon the subject.

The questions upon which we have attempted to throw light are as follows:

1. The relative value of vegetable as compared with animal foods as sources of the albuminoids (nitrogenous or flesh forming part) of food.

2. A comparison of dried animal or flesh meal with fresh cut bone as foods for laying hens.

3. The value of condition powders as an ingredient in the food of hens.

4. The influence of the presence of a cock with the hens upon the production of eggs.

An earnest effort has been made in planning and carrying out these experiments to eliminate all disturbing influences—to insure, in short, in every instance perfect fairness, perfect equality in all conditions other than the one variation in feed or treatment which constituted the subject of experiment. It is to be feared that in many of the private experiments which have been carried out such equality in conditions has not always been secured. The one feed or treatment is tried during one part of the year—another during some other period—or perhaps even, one method of feeding has been the subject of experiment one year; another method some later year. It must be evident that under these conditions results cannot fairly be compared. Every variation, save the one under trial, must be eliminated if results are to have much value.

Attention is called to this matter—not as a reproach to the practical man; he is rather entitled to all honor for the many valuable discoveries he has made. Under the conditions under which he works, under the urgent necessity usually existing to make his business financially successful, it is rather to be wondered that his experiments have been even so well carried out as has been the case. No! not then, in criticism of the practical man, but rather to explain why it is that exact knowledge is still wanting upon many questions of great importance, is attention called to the matter.

In what measure we have secured suitable conditions for just comparison in our experiments will be apparent from the brief statement which follows:

1. Houses—In every instance where two pens of fowls have been compared they have occupied separate detached houses precisely similar in every dimension and detail of construction and with the same aspect and exposure. Each house has two rooms, a roosting and laying room, 10 by 12 feet, and a scratching shed, 8 by 10 feet in size. The former has two windows of the ordinary size; the latter is open to the south in all but the most extremely cold or stormy weather. It may be closed by folding doors, each with glass windows. In this shed is scattered eight or ten inches of cut straw. With each house is connected a yard containing 1,200 square feet, and of this the fowls have always had free run.

64—WHITE COCHIN BANTAM CHICKS

2. Care and Cleanliness—The utmost regularity in care and feeding has always been observed in every particular; droppings have always been promptly removed and suitable measures taken to keep down vermin.

3. The Fowls—In every experiment the fowls in the two coops under comparison have been of the same breed and age; and of the same origin and past treatment in each coop. In preparing for experiments we first place all the fowls together, then take at random one for the first coop, next a fowl as nearly like the first in every particular as can be found in the lot for the second coop; and so on to the end. In other words, the fowls are matched in pairs as closely as possible and one of each pair goes into each of the two pens which are to be compared. In most cases we have raised our experiment stock; but if some instances we have purchased it. We have used fowls of the following breeds: Light Brahma, Barred Plymouth Rock, White Wyandotte and Black Minorca. We have not attempted a comparison of breeds.

4. Feeding—Our custom has been to feed a warm mash early in the morning; to scatter some whole grain in the straw of the scratching shed at noon, and to give more whole grain also in the straw of the shed about an hour before sunset. The aim has been to give all that the fowls would eat with a relish; but to make them work for the whole grain. The quantities fed have been determined by the judgment of the feeder; but an exact account of kinds and amounts of feeds has been kept and these feeds have been subject to analysis. They have been sound and of good average quality. Artificial grit, shells and pure water have been kept always at hand.

The opinion is ordinarily held that in order to procure a satisfactory egg yield hens must be given some animal food, and this is stated to be necessary in order that the ration may be sufficiently nitrogenous. There are vegetable substances now available, however, which are so rich in albuminoids that a ration nitrogenous enough to meet all requirements can be made up by their use.

Among these the soy bean in the one case and cotton seed and linseed meals in the other were selected for comparison with meat meal in the two experiments which we carried out upon this question in 1895.

The soy bean was ground into a fine meal, which is even richer than the ordinary meat meals, as is shown by the table:

Soy bean meal, moisture.......................11.61 per cent
Meat meal, moisture ....13.68 per cent

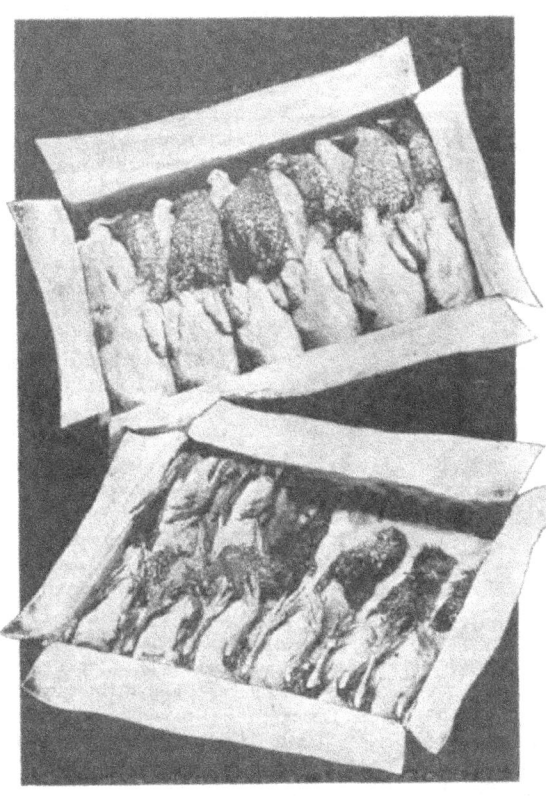

65—CRATE-FED CHICKENS IN CASES

COMPOSITION OF DRY MATTER, SOY BEAN MEAL AND MEAT MEAL.

	Albuminoids	Fat.	Carbohydrates.
Soy bean meal, per cent.....	34.37	16.38	45.22
Meat meal, per cent..........	35.98	8.31	0.00

In both experiments the fowls received a variety of other foods, but the nutritive ratio was kept substantially the same for the two coops under comparison. The foods used in the first experiment in addition to the soy bean meal were cut alfalfa, wheat, oats, and middlings in the one coop; in the other, boiled potatoes, ground clover, wheat, wheat middlings and cut bone.

In the second experiment the supplementary feeds were wheat, oats, bran and middlings for the vegetable coop and wheat, oats, wheat meal, bran and linseed meal for the animal food coop.

The number of fowls in a coop was twenty. The first experiment lasted sixty-four days, December 9th to February 12th. The other one hundred and fifty-three days, January 1st to October 31st. The results were decidedly favorable to the animal food. The egg yields, however, were small in both cases.

ANIMAL MEAL VERSUS CUT BONE

We have now carried through the five experiments comparing the dry flesh meal with fresh cut bone. At the outset our practice was to give the bone by itself. This practice we soon gave up as it was very uneven and not infrequently individuals obtained sufficient to purge them seriously. In all these experiments we have used a variety of foods and have endeavored to keep the nutritive ratio substantially the same in the two coops under comparison.

Two of our experiments have given results slightly favorable to the bone in number of eggs; one a result slightly favorable to animal meal; and two, the last two, which have been most carefully carried out, have given results most decisively favorable to the animal meal. The latter has invariably been found the safer food even when the bone is fed in the mash. Some fowls invariably scratch and obtain more than their share. Such fowls suffer from diarrhoea, which sometimes proves fatal, and always, of course, lessens the egg production for the time being.

To give a clearer idea of the method of experiment, details concerning the last one will be given.

This experiment continued from December 12th, 1897, to April 30, 1898. There were nineteen Barred Plymouth Rock pullets in each house when the experiment began. Those in the flesh meal house weighed 101.5 pounds and had laid, November 8th to December 12th, eighty-two eggs. The pullets in the cut bone house weighed 101.25 pounds and had laid forty-one eggs. In the morning mash of one lot one part of animal meal to five parts total dry material was used. In the mash of the other lot the same proportion of fresh cut bones was mixed. The large, flat bones, comparatively free from meat or fat were used. In the animal meal coop the health of the fowls was good, but one fowl being out of condition in any way. The nature of the trouble was unknown. The fowl was killed.

Almost from the first bowel troubles were common in the cut bone coop. Two fowls died, December 23d and January 11th. One hen met with an accident and was killed.

The hens in the animal meal coop laid three soft shelled eggs; the others two.

The bone fed amounted to only .27 ounces per hen daily; .5 ounces and over is usually recommended by writers upon the subject. We find it impossible to feed so largely without serious bowel trouble.

The tables will give a clear understanding of the experiment and its results.

FOODS CONSUMED

Kind of Food.	Pen 1. Animal Meal. lbs. oz.		Pen 2. Cut Bone. lbs. oz.	
Wheat	256	0	262	0
Oats	143	0	145	0
Bran	44	8	39	0
Wheat middlings	44	8	39	0
Gluten feed	44	8		
Gluten meal			39	0
Animal meal	44	8		
Cut bone			40	0
Clover rowen	44	8	39	0
Cabbage	19	3	18	8

AVERAGE WEIGHTS OF THE FOWLS (POUNDS)

Dates.	Animal Meal.	Cut Bone.
December 12	5.34	5.38
January 31	5.64	5.66
February 25	5.66	5.88
March 30	5.09	5.27
April 30	5.06	5.53

EXPERIMENTS—LAYERS AND EGGS

EGGS PER MONTH (NUMBER)

Months.	Animal Meal.	Cut Bone.
December	63	57
January	92	83
February	184	120
March	263	259
April	210	209
Total	812	728

ANIMAL MEAL VS. CUT BONE FOR EGG PRODUCTION

	Animal Meal.	Cut Bone.
Total number of eggs	812	738
Hen days	2,561	2,331
Gross cost of foods	$8.45	$8.29
Cost per egg	$0.0104	$0.0114
Cost per hen day	$0.0033	$0.0035
Total weight of eggs (pounds)	100.5	88.7
Average weight per egg (ounces)	1.98	1.95
Eggs per hen day	.32	.31
Dry matter consumed per hen day (lbs)	.22	.23
Dry matter to produce one egg (lbs)	.695	.739
Nutritive ratio	1:4.6	1:4.7
Sitters	22	13

It will be noticed that the fowls receiving animal meal laid many more eggs of greater average weight than those receiving the bone. The cost per egg for food was considerably less. The animal meal is, moreover, a more convenient feed to use as well as safer. The fowls at the close of the experiment weighed less, it is true, where animal meal had been the food, but the slight loss in weight is far more than covered by the greater value of the egg product.

A test of the eggs, both raw and boiled, was made by an expert, who pronounced the animal eggs somewhat inferior in color and flavor to the others. The experiment carried out during the previous winter gave results even more decisive in favor of the animal meal.

As a result then of a most painstaking and long continued inquiry upon this subject I am convinced that the dry fine animal or flesh meals, if sweet and good quality, are much to be preferred as a source of animal food to fresh cut bone.

CONDITION POWDERS

There seems to be a quite widespread opinion that something hot, something of the nature of a condiment, mixed with the mash given to laying fowls is useful. This idea receives encouragement also from some of the most prominent poultry papers and writers. An honest desire to know whether such condiments are needed or useful led to the investigations the results of which are here given.

Three experiments have been carried out. We have used in all one of the most generally advertised and recommended of the various condition powders. These powders have been used in accordance with the directions sent out by their makers. The difference in number of eggs with and without the powder has in every case been small. In one experiment a few more eggs were laid by fowls receiving the condition powder; in two experiments those not receiving the powder laid more eggs.

In the last experiment twenty Barred Plymouth Rock pullets were used in each coop. The experiment continued from December 12th to April 30th. The fowls in both coops received a variety of foods, including wheat, oats, bran, middlings, gluten feed, animal meal, cut clover and cabbage, both kinds and amounts being practically the same as in the experiment comparing animal meal with cut bone.

The leading results of the experiments are shown by the tables:

EGGS PER MONTH (NUMBER)

	Condition Powder.	No Condition Powder.
December	28	59
January	90	66
February	86	101
March	217	288
April	298	291
Totals	719	745

CONDITION POWDER FOR EGG PRODUCTION
(December 12th to April 30.)

	Condition Powder	No Condition Powder
Hen day	2,751	2,656
Gross cost of food	$8.91	$8.51
Cost per hen day	$0.0032	$0.0032
Total number of eggs	719	745
Cost per egg, not including powder	$0.0124	$0.0115
Cost per egg, including powder	0.0180	0.0115
Eggs per hen day	.26	.28
Total weight of eggs (pounds)	88.08	90.80
Average weight of eggs (ounces)	1.96	1.95
Dry matter to produce one egg (lbs)	.82	.77
Dry matter consumed per hen day (lbs)	.22	.22
Nutritive ratio	1:4.6	1:4.6
Sitters	8	14

66—PACKED AND READY FOR SHIPMENT

There was no noticeable difference in the health and condition of the fowls in the two coops during any part of the experiment. Eggs from both lots of fowls were tested under numbers by two families. One reports no difference in quality. The other found the eggs from the ones not receiving condition powders "much preferable" to the others.

As a result of our experiment I conclude that the use of condition powders is not beneficial. The differences have never been sufficiently great to be significant. In two cases the hens not getting the powder have produced more; in one case the others have produced a few more. In the light of

EGGS AND EGG FARMS

these results it is believed that poultry keepers throw away money expended for condition powders.

INFLUENCE OF MALE BIRD

Certain experiments which have been carried on at some of our Experiment Stations have given results indicating that hens will lay more eggs when kept by themselves than they will if a male is kept with them. Should the effect of a male be found to be an invariable decrease in the number of eggs the fact would be of considerable importance as affecting the economy of the production of eggs for market. It was accordingly deemed advisable to try this experiment. Accordingly four coops of fowls (sixteen in each) were selected and cocks placed in two of them. The experiment continued from May 13th to September 2nd. As will be seen there were really two experiments. The results indicate that the cock was apparently without influence upon the number of eggs. In one pair of coops the fowls with the cock laid 631 eggs; those without a cock, 630. In the other pair of coops the numbers were respectively 629 and 526. Exact comparison is possible, however, only when we ascertain the product per hen day. This is found to be for the first experiment .35 eggs for the hens with a cock and .36 for those without. In the second experiment it is found to be .38 for the hens with a cock and .33 for those without.

It will be noticed that in the one case the hens with which the cock was kept make the best record, while in the other case those kept without a cock make the best record. The differences are small and since they are in the one case on one side and in the other case on the other, the conclusion appears inevitable that the male was without influence on the number of eggs produced.

What then are the hints on feeding that I would base on our experiments? Briefly, they are these:

1. Some animal food is essential to satisfactory egg production. Vegetable foods, even equally rich in the nitrogenous and fatty ingredients, are much inferior.

2. The dried animal or flesh meals made a part of the morning mash are safer and better animal food than cut bone; and they cost less and are more convenient in use. The proper proportion appears to be about one part by weight of the animal meal to five parts of the other dry materials used in the mash.

3. Condition powders (and presumably other condiments, such as common and cayenne peppers) are useless and money spent for them is thrown away.

5. The male bird exerts no appreciable influence on the number of eggs produced.

67—A SOFT ROASTER

GREEN BONE AND MEAT MEAL

WEST VIRGINIA EXPERIMENT STATION COMPARES THE VALUE OF MEAT MEAL AND GROUND FRESH MEAT AND BONE AS EGG PRODUCING FOODS

IT IS well known that the ground fresh meat and bone is a very valuable constituent of a ration for egg production. In many localities, however, it is difficult to procure fresh bones and scraps from meat markets, and even when a supply is constantly available it is not usually an easy matter to grind the material for the fowls. On the other hand, beef scraps and meat meal can be bought of the poultry supply houses at any time, and being in a thoroughly dry condition, can be readily mixed with other feeding stuffs.

The experiment described below was undertaken for the purpose of comparing meat meal with ground fresh meat and bones as materials furnishing protein to laying hens.

Thirty-four Barred Plymouth Rock hens and two cocks were divided into two similiar lots. They were supplied at all times with mica crystal grit, granulated bone and water. The grain ration for each lot of fowls was the same, but the amount consumed varied somewhat, and so the actual amount of grain consumed by each lot is given. At the beginning of each period the grain for that period was weighed and stored in suitable boxes. No account was taken of the daily amounts fed. At the end of each period the amount remaining was again weighed, and the difference taken as the actual amount of food consumed.

The experiment began October 26, 1899, and was continued for four periods of thirty days each.

During the experiment the fowls receiving the fresh bone laid 3,824 eggs weighing 495.2 pounds, of an average weight of 12.75 pounds per hundred eggs, while the meat meal lot laid only 3,260 eggs weighing 391.2 pounds and weighing 11.94 pounds per hundred. Consequently the fowls fed fresh bone not only gained more in weight, but they also laid more and larger eggs.

[NOTE.—The results of this experiment with green cut bone and meat meal seems to contradict the conclusions drawn by Prof. Brooks from similiar experiments made at the Hatch Experiment Station. Undoubtedly further tests of the matter will be made and results duly reported. Most poultrymen advise the liberal use of green cut bone provided always that it is from suitable animals. If the food that is the easier to obtain does not agree with your fowls after a thorough trial, it will pay to buy the other and you may do so with the satisfaction of knowing that the second choice has been tried with excellent results.—ED.]

EXPERIMENTS—LAYERS AND EGGS

PRODUCING AN EARLY MOLT

AN EFFECTIVE METHOD OF FEEDING TO HASTEN THE MOLT AND SO PREPARE FOWLS FOR WINTER LAYING

J. H. STEWART AND HORACE ATWOOD

WHEN a specialty is made of producing winter eggs it is of much importance to have the hens shed their feathers early in the fall so that the new plumage may be grown before cold weather begins. In case molting is much delayed the production of the new coat of

68—VIEWS AT AN ENGLISH POULTRY FARM

feathers in cold weather is such a drain on the vitality of the fowls that few if any eggs are produced until spring, while if the molt takes place early in the season the fowls begin winter in good condition and with proper housing and feeding may be made to lay during the entire winter.

A few years ago Mr. Henry Van Dreser proposed a way whereby fowls may be caused to molt as early in the fall as is desirable. Briefly this method consists in withholding food either wholly or in part for a few days, which stops the egg production and reduces the weight of the fowls, and then feeding heavily on a ration suitable for the formation of the feathers and the general building up of the system.

The experiment designed to study this method was begun Aug. 5; with two pens of Rhode Island Reds, and two pens of White Leghorns, about two years old. One pen each of Rhode Island Reds and White Leghorns received no food for thirteen days except what they could pick up in their runs, which had been sown to oats in the spring. These runs are fifteen feet wide and one hundred feet long and nearly all the oats had been picked from the heads before the beginning of the experiment. The other two lots of fowls were fed as usual on mash, beef scraps, corn, wheat, and oats. After the expiration of the thirteen days all four lots of fowls were fed liberally. Each lot of fowls contained twenty hens and two cocks. The following table shows the number of eggs produced during the first thirty days after the beginning of the test:

LOT	BREED	HOW FED	EGGS PRODUCED
1	Rhode Island Reds	Fed Continuously	75
2	Rhode Island Reds	No Food	17
3	White Leghorns	Fed Continuously	172
4	White Leghorns	No Food	25

Lots two and four ceased laying entirely on the seventh day of the test.

Thirty days after the test began the "no food" pen of Rhode Island Reds had practically a complete coat of new feathers, had begun to lay, and within a week from that time one-half of the hens were laying regularly, while the other lot of Rhode Island Reds were just beginning to molt, and the egg production had dropped down to two or three eggs per day. Both lots of White Leghorns were a trifle slower in molting than the Rhode Island Reds, but otherwise the treatment affected them in a similar way. For ten days, beginning August 19th, the droppings boards in the two White Leghorn houses were not cleaned.

SUMMARY

Mature hens, which are fed very sparingly for about two weeks and then receive a rich nitrogenous ration, molt more rapidly and with more uniformity, and enter the cold weather of winter in better condition than fowls fed continually during the molting period on an egg producing ration.

WEIGHT OF EGGS IN INCUBATION

TESTS BY THE WEST VIRGINIA STATION TO DISCOVER LOSS OF WEIGHT IN EGGS DURING INCUBATION

A RECENT bulletin of the West Virginia University Experiment Station, Morgantown, W. Va., is devoted to a report of loss in weight of eggs during incubation. Details of three tests are given, showing the original weight of the eggs, their weights respectively at the end of the fifth, twelfth and nineteenth days, and a comparison of the weights of infertile eggs with those that hatched, and with others that contained chicks dead in the shell. The summary and conclusion give results of the experiments as follows:

1—Fertile eggs, when incubated in normal manner, decrease in weight.

2—The eggs which hatched lost 4.17 per cent. of their weight during the first five days of incubation. During the seven succeeding days they lost 6.35 per cent. of the weight of the eggs at the end of the fifth day, and during the next seven days lost 6.98 per cent. of their weight at the end of the twelfth day.

3—One hundred fertile eggs of average size will lose 234.9 grams, or 8.28 ounces, during the first five days of incubation; 341.8 grams, or 12.05 ounces, during the next seven days; and 352.8 grams, or 12.44 ounces, during the next seven days.

4—The infertile eggs lost 3.6 per cent of their original weight during the first five days of incubation. During the seven succeeding days they lost 5.6 per cent of what they weighed at the end of the fifth day, and during the next seven days lost 5.6 per cent of their weight on the twelfth day. One hundred infertile eggs will lose 217.2 grams, or 7.66 ounces, during the first five days; 323.3 grams, or 11.40 ounces during the next seven days; 306.9 grams, or 10 82 ounces, during the next seven days.

PULLETS VS. HENS AS LAYERS

RESULTS OF EXPERIMENTS SHOWING THE RELATIVE VALUE OF HENS AND PULLETS FOR EGG PRODUCTION

JAMES DRYDEN

PROBABLY no question is of more importance in commercial poultry keeping than that of the value of the hen at different ages. Successful poultrymen, of course, understand that the hen does not improve with the keeping or with the training, as some men do. They fully realize that they can get no profit out of their business unless they renew their flock every two years or at most every three years, but it is safe to s y that the vast majority of the hens of the country have reached the unprofitable age, or at any rate are of doubtful age, and it would be millions of dollars in the pockets of the keepers of poultry in the United States if they would at once kill off and market every hen that had not an unmistakable mark of being a spring chicken, and start new. But most people ask for the evidence.

It has been shown by a series of experiments at the Utah Station that the young hen is the more profitable hen and that the old hen is not profitable. The first and second years showed a good margin of profit, but the third and fourth years little or none. In one series of experiments the record showed the yearly egg yield of two seperate pens of Leghorns for three years, and of two pens for two years. The following table gives the record of eggs laid in full:

PEN	NUMBER OF EGGS LAID PER HEN		
	1st Year	2d Year	3d Year
3—Fed all grain in box	157¼	150½	134
4—Fed all grain in straw	181½	114½	99
1—Fed all grain in box	160¼	119	
2—Fed all grain in straw	157	120	
Average	164	126	116

Taking the average of the four pens it is seen that over thirty per cent more eggs were laid the first year than the second and fifty per cent more than the third. Accepting these results as fairly representing average conditions no one will pretend to say that hens should be kept three years, but there may be an honest difference of opinion as to the advisability of keeping hens two years. There is no possible reason to doubt that the first year is the most profitable. If the eggs laid were sold at two cents a piece, or 24 cents a dozen, the first year's eggs would bring 76 cents more than the second year's eggs.

From here on we have no experimental evidence, so we must do some figuring, we must make some deductions. The way it should be settled is this: A farmer has one hundred hens that have just finished their first year's laying, say in August or September. The question that concerns him is: Shall I keep those hens another year or shall I market them and buy pullets (in case he has not raised pullets to take their places)? Knowing the price he can get for his hens and the price he would have to pay for pullets the problem would be easily solved. If he gets 25 cents for the hen, and pullets eggs are worth 76 cents more than year old-hens' then he has $1.01 with which to buy the pullet, and he can probably buy pullets for fifty cents each or hatch and raise them for less. The latter proposition, however, is the question in doubt—the cost of hatching and raising the pullet, a question which has never been demonstrated, but the evidence is in favor of using only pullets for laying eggs for market.

Since the above results were published, records of a number of individual hens of different breeds have been completed for two years and in some cases three and four years. These records show that in exceptional cases more eggs are laid the second year than the first, though the average of all individual records bears out the former conclusions that the first year is the most profitable year. They show that there is a question of individuality involved. Hen No. 131, for instance, laid 241 eggs the second year and 201 the first. Hen No. 155

69—THE TOLMAN HOUSE IN WINTER

laid 197 the second year and 186 the first; hen No. 68 laid 165 the second year and 183 the first.

Averaging the results of all individual hens for which we have two years' records, we have the following: Thirteen Brown Leghorns laid an average of 193 eggs the first year per hen and 157 the second, as shown in table No. 32; four White Leghorns averaged 183 the first year and 95 the second; five Barred Plymouth Rocks averaged 154 eggs the first year and 110 the second; sixteen White Wyandottes averaged 170 the first year and 111 the second; three White Plymouth Rocks averaged 192 the first year and 129 the second.

There are forty-one hens of the different breeds for which individual records are complete for two years, and we find that they averaged 178 eggs per fowl the first year and 125 the second, or forty per cent more the first year than the second. These results refer to yearly records. The question whether the hen or the pullet is the better winter layer, is of importance.

The forty-one individual hens whose records were kept, laid 916 eggs as pullets before February 1st, and as year-old hens they laid only 437 to the same date. This would show that the pullets not only lay more eggs during the year than the hens, but they a larger proportion of them in winter than the hens.

EXPERIMENT IN FEEDING FOR EGGS

RATIONS THAT WERE FED TO ENFORCE EGG PRODUCTION DURING A COMPETITION WHICH RESULTED IN AN AVERAGE OF TWO HUNDRED AND EIGHTY-NINE EGGS PER HEN PER ANNUM IN THE WINNING PEN — NUMEROUS HELPFUL SUGGESTIONS

IN the egg contest promoted by the National Stockman and Farmer, Pennsylvania, 224 pens of fowls were entered. Weekly reports were required from each contestant, and the value of the eggs laid was determined according to the current price of eggs in the Pittsburg market, this value being computed on the number of eggs as reported from week to week. The six highest winners and the number and value of eggs were reported in the Stockman and Farmer as follows:

First—Pen 112, W. S. Stevens, Ohio, eight White Plymouth Rock pullets, an average of 289 eggs each, or a value of $5.02 per hen.

Second—Pen 189, William G. Dodson, Ohio, eight crossbred Leghorn pullets, an average of 283 eggs each, or a value of $4.82 per hen.

Third—Pen 115, J. G. Redkey, Ohio, eight White Plymouth Rock pullets, an average of 280 eggs each, or a value of $4.00 per hen.

Fourth—Pen 75, L. E. Bradbury, Ohio, eight Single Comb Brown Leghorn Pullets, an average of 277 eggs each, or a value of $4.64 per hen.

Fifth—Pen 88, Z. N. Allen, Pennsylvania, twenty-four Single Comb Brown Leghorns, an average of 277 eggs each, or a value of $4.89 per hen.

Sixth—Pen 154, Z. N. Allen, Pennsylvania, twelve Barred Plymouth Rocks, an average of 262 eggs each, or a value of $4.24 per hen.

Not being satisfied with a mere knowledge that a stated number of fowls laid a stated number of eggs, an effort was made to get down to hard pan by securing from the winners the actual facts. They are given in the words of the contestants.

AN AVERAGE OF TWO HUNDRED AND EIGHTY-NINE EGGS PER HEN

"You ask how I managed and cared for my eight White Plymouth Rock hens during the recent egg contest as conducted by the National Stockman. I will be pleased to tell you. This pen consisted of eight White Plymouth Rock hens and one rooster. These eight hens laid 2,312 eggs in 365 days, or an average of 289 per hen for the year. Estimated by the Pittsburg (Pa.) market, week by week, each hen laid during the year $5.02 worth of eggs. They were kept in a house 12 by 20 feet long, divided into two parts, each 10 by 12, one part being used for a scratching shed and the other part containing the nests and roosts. The building is seven feet high and is a frame, weatherboarded with pine siding and ceiled with matched pine flooring, which makes the house very warm. You will notice this pen had plenty of room. The floor consists of mother earth and is covered about four inches deep, in the fall, with road dust and sand. The building runs east and west, facing the south.

"In the south of the building are two windows, which extend from the floor to the height of the building, thus admitting plenty of sunshine and light, so necessary to the comfort and happiness of the fowls. The perches are about three feet from the floor, and under them the dopping boards. A house of this kind in which fowls are housed during the winter months, with the right kind of food and the proper care, will insure the poultryman eggs all winter. My hens were not out during last December and January, and they were as healthy, happy and contented as if they were roaming the fields during the summer months. They were all aglee with song and contentment and shelled out eggs every day, even during the coldest days of last winter.

"They have free access to oyster shells and grit. I give them twice a week fresh granulated bone. Their food consists of a warm breakfast, equal parts of bran, white middlings and chopped corn and oats, and into this I put for them fine beef loaf. At noon I feed wheat, which is thrown into the scratching shed. This gives them exercise in obtaining their noon meal. In the evening they are fed whole corn. During the time from the first of April until the first of November, I fed the same, with this change: In the morning their mash is mixed with cold water; in the evening wheat takes the place of corn. Cleanliness is a very important matter in regard to to the maintenance of health for your fowls. I clean the house twice a week during the winter and in the summer every other day. I have been breeding Plymouth Rocks now for five years, and have as yet not had any disease, and I attribute it to cleanliness and proper care."

W. S. STEVENS.

WINNER OF SECOND PRIZE

Mr. William G. Dodson, who won the second prize, wrote as follows:

"My pen of eight pullets that I had in the National Stockman and Farmer contest laid an average of 283 eggs each in one year. The pullets were from a Rose Comb Brown Leghorn cock crossed by White Leghorn hens. The pullets I had in the contest were the result of that cross. The house I kept them in was built of lapsiding and lined with Neponset paper and roofed with the same. Not a pin crack was left for drafts to get in. I have a good-sized yard fenced in wire netting. Every morning these pullets had a hot feed of chop, mixed with the water that the fresh bones and beef scraps were boiled in. After that some wheat and oats were thrown in the straw for them to scratch for. At noon they had ground bone and meat scraps and stale bread. At night they had in summer wheat and barley, and in winter corn and buckwheat, and at all times they had before them fresh water, and each day fresh milk. Twice a week I gave them some buttermilk. They also had at all times a good supply of broken dishes, seashells and limestone, broken in small pieces, and once a week they had a small quantity of ground ginger and black antimony.

"The house was cleaned once a week and the floor sprinkled with air-slacked lime, and the inside of the house dosed with coal oil. The dust box was four feet square and filled with sifted coal ashes and road dust mixed. Not one of them was sick or "off its feed" one hour in the whole year, and they are still laying and look as fresh as any of my chicks. They are my best layers singled out for several years, I breed from none but the best. I have been experimenting for some time in crossing different chicks. I could just as well have entered a pen of full bloods as cross breeds, but so many laughed at cross breeds I thought I would give them a trial."

WILLIAM G. DODSON.

EGGS AND EGG FARMS

AN INTERESTING COMMUNICATION.

Mr. J. G. Redkey, won the third prize with eight White Plymouth Rock pullets that averaged 280 eggs each for the year. He wrote as follows:

"The varieties I breed are pure-bred White and Barred Plymouth Rocks. I feed warm food in the morning, composed of cooked meat two parts and twenty parts of cracked wheat, with whole wheat and oats at noon scattered in litter. I feed oats, wheat and corn at night, with clover heads, cabbage, beets or turnips for green food, and cut bone, oyster shells and crushed limestone for grit.

"My houses are built 14 by 20 feet, with a hall 4 feet wide in front and four six-light windows in front. There is a partition in the center, making two pens 10 by 10 feet to each house. These houses are double boarded, with tarred paper between, and are roofed with Marietta roofing, double seamed. Each house is five feet high in the rear and eight feet in front.

70—LIGHT BRAHMA BREEDERS

Each house has an earth floor filled in with from six to eight inches of pounded clay, with four inches of coal cinders on top, which makes a floor perfectly dry.

"These houses are frost-proof, having withstood a temperature of twenty-one degrees below zero: This, I think, is one of the great secrets of winter egg production, as my twelve years' experience as a breeder of pure-bred poultry has taught me that you can not expect to get eggs in winter with all the feeding and care you may be able to give unless you have comfortable houses for them.

"There is also a great difference in the laying qualities of birds of the same breed, some strains laying almost double the number of eggs of others of the same breed. I have been mating some of my pens with that object in view, viz: eggs, and I have been in a measure successful, as my record in the late contest shows. I have been giving this my attention for the past eight years, and by careful selection have increased the average per hen from 212 eggs nine years ago to 280 in 1894. In my pens of White Plymouth Rocks and in Barred Rocks I have brought them up from 205 to 264 in the same length of time. My yards are each 30 feet wide by 200 feet long, with one house for each two yards. Each pen contains fifteen hens and one cock, except the pens that were in the contest, which contained nine hens and one cock, and ten hens and one cock respectively.

"I have never allowed my hens to rear chicks, as I hatch and rear all my fowls by artificial heat, and when I have a hen that becomes broody I remove her to a yard prepared for that purpose, containing no nests or secluded corners, and in a few days she can be returned to the pen, and will soon be laying again, as though she had never offered to sit. It is my belief that fowls hatched in incubators and reared in brooders year after year will lose, to some extent, the habit of incubation, as my Rocks are now much less inclined to become broody than they were a few years ago, and I firmly believe that were it possible to introduce no other blood in the yearly matings, except from those that were artificially hatched and reared, the results would be much more marked. I may be wrong, but I had in one of my pens a Barred hen hatched May, 1893, that laid 297 eggs to March 1, 1895, and never offered to sit. This is an exception, but only goes to prove what I believe is possible."

J. G. REDKEY.

NOTE—Mr. Redkey is confirmed in his opinion by the report of the United States Consul to Egypt, which states that the native hen of that country (where artificial incubation has been followed for centuries) has abandoned the work of hatching.

NO EXCELLENCE WITHOUT LABOR

Mr. Z. N. Allen, who came off fifth and sixth best in the contest, receiving an average of 277 eggs each from a pen of twenty-four Single Comb Brown Leghorns, and an average of 262 eggs each from a pen of twelve Barred Plymouth Rocks, favors us with the following valuable information:

"Sixteen years ago this spring I began an egg contest of my own. The preceding summer I had built a good hen house, so I determined to ascertain which of my breeds were the best egg producers. I penned six of each kind, Brown Leghorns, Silver Hamburgs, Polish and Plymouth Rocks. This gave me some experience in feeding and confinement (which lasted four months) and this experience has stood me well in hand ever since. Pens Nos. 88 and 154 in The Stockman egg contest were pullets from good laying stock. Those in No. 88 were hatched the first week in May, 1893; began laying about the middle of November. Those in Pen No. 154 were hatched the first of April, 1893, and began laying the last of November. They were well fed and cared for from chicks until the contest ended. Their houses were made as warm as could be without artificial heat. Their apartments were kept clean and dry and were supplied with grit, ground bone and oyster shells. They had to scratch in winter in litter, and in summer in sand. One side of their yards was spaded two feet wide. Then wheat was scattered and the sand was shoveled up against the side of the yard. To get the wheat they had to scratch it back until it was about level. This was repeated once a day during the summer unless it was too wet. When cold weather came they had to go into winter quarters and earn their living by scratching litter. They breakfasted on hot mash in winter and not very cold in summer. A short time after breakfast they went to scratching for life, some singing as they worked. For dinner they had green bone, meat and clover every alternate day, with very little exception. The noisiest time in the hen house was from daylight until noon. I thought sometimes they were trying to see which could make the most noise. I believe they had a more jolly time than I did. Along in the afternoon they turned their scratching into pecking cabbage. This sobered them down somewhat and gave them an appetite for supper, at least they got in a hurry and tried to see which would get the first bite. They appeared to relish very much boiled wheat and oats and some coarse bran, even if it was quite hot. Once a week the medicine man came around with his jar of cayenne pepper and made it a little hot for them in the mash. I believe the pullets in 88 and 154 laid eggs because they liked to do it. Pen 88 made a record of 6,654 eggs; Pen 154, 3,139 eggs. Pure-bred stock, good, wholesome food and plenty of it, good warm houses and good care will make a success of poultry."

Z. N. ALLEN.

AN EGG RATION

THE MONTANA EXPERIMENT STATION CONDUCTS EXPERIMENTS WITH A VIEW TO REDUCING THE COST OF EGG PRODUCTION—VALUE OF VEGETABLES, MEAT AND GRAIN

IN MARCH, April and May, 1900, we fed four pens of fowls, sixteen in each, fifteen hens and one cock, upon four different rations, with a view of determining what effect a variety ration (meat, vegetables and grain), a meat ration (meat, meal and grain), a vegetable ration (vegetables, meal and grain), and a straight grain ration had upon egg production.

The fowls were housed in a log building in pens 9 by 10 feet with yards 10 by 16 feet. The yards were very small, as the ground in front of the building was being graded. The fowls obtained no vegetable food whatever from the yards, as they were covered with chaff and straw. Grit, burnt bone and dust baths were supplied the fowls alike, as was plenty of fresh water. The birds all remained quite healthy throughout the duration of the experiment, two months and a half, no loss or any disease occurring. Each pen contained eleven two-year-old hens and four pullets and one cock. The male birds were all vigorous yearling Plymouth Rocks and well developed, so that a maximum amount of service could be expected from each of them. The hens were about one-half scrub stock—Cochin, Game, Leghorn and Rock mongrels, and the remainder pure-bred Plymouth Rock hens. They were as evenly divided in respect to variety as possible. Feeding was done three times a day—about 7 A. M., again at 11:30 A. M., and from 4 to 4:30 P. M. In each pen the floor was covered with litter and the grain fed therein, so that, though closely confined, all had plenty of exercise.

RATIONS

Pen No. 1 received in the morning 12-ounce feed, ¼ bran, ¼ oat chop, ¼ meat, ¼ vegetable, then a mangold was given, and at noon clover with a little meat or ground green bone. Evening feed was grain (wheat or oats).

Pen No. 2 received in the morning 12-ounce feed, ¼ bran, ¼ chop, ½ meat, later some grain, and at noon a little meat or bone. Evening feed was grain (wheat or oats).

Pen No. 3 received in the morning 12-ounce feed, ¼ bran, ¼ chop, ½ vegetable, then mangold and a little grain, and at noon clover and roots, and in evening grain (wheat or oats.)

Pen No. 4 received in the morning 12-ounce feed of chop, ⅓ bran, ⅓ oats, mixed with warm water.

COST OF DIFFERENT RATIONS

Pen No. 1—Cost of bran, oat chop, meat and bone, vegetable, grain, $1 97.

Pen No. 2—Cost of bran, oat chop, meat and bone, grain, $2.03.

Pen No. 3—Cost of bran, oat chop, vegetable, grain, $1.79.

Pen No. 4—Cost of bran, oat chop, grain, $1.95.

It was the endeavor in composing these different rations to show the advantage of a variety, and the variety of feeds used was such as could be made use of generally.

In the second ration we endeavored to show the value of a succulent vegetable feed by eliminating it.

In the third the value of meat and bone was demonstrated in the same manner.

In the fourth we endeavor to show the fallacy of feeding, as many do, a straight grain ration. And in the results it was shown, from egg production, that the greatest egg yield was received from the first, the variety ration, while the smallest returns came from the hens fed upon the grain alone. The advantage here must be very apparent, since the cost of both rations was almost the same. The following table shows briefly and concisely the egg yield from the different pens, the weight of eggs, and their market value:

Pen	No. Laid	Weight	Cost	Value	Gain
1.................	431	45-8 ounces	$1.97	$8.98	$7.14
2.................	407	44-4 ounces	2.03	8.48	6.58
3.................	366	39-11 ounces	1.79	7.62	5.94
4.................	342	36-6 ounces	1.95	7.12	5.29

71—SPECIAL DUST BOX

In computing the cost of food in this experiment the following values were used.

	Per cwt.
Oat chop	$ .96
Bran	.70
Oats	.90
Wheat (frosted)	.40
Mangolds	.75
Clover	.30
Potatoes	.50
Beef and Bone	1.00

The eggs were valued at 25c a dozen.

The financial results of this experiment are excellent. It has been shown that even where fed upon grain alone and closely confined, considerable gain was made, while the pen receiving the greatest variety of food, costing about the same, made a further gain of $1.84. The total returns of pen No. 1, deducting the cost of food, being $7.13 and from pen 4 being $5.29. In the meat and vegetable fed pens the one receiving the meat and bone, though the more expensive ration, was still more profitable, yielding a profit of 64 cents over pen No. 3, the total profits from the four pens of 60 birds being $24.94, or 41½c per bird for period of experiment—two and one-half months.

PRESERVATION OF EGGS

THE RHODE ISLAND EXPERIMENT STATION MAKES TESTS OF VARIOUS METHODS OF PRESERVING EGGS AND SECURES RESULTS THAT WILL BE OF INTEREST TO POULTRYMEN—DETAILS OF EXPERIMENTS—WATER GLASS METHOD PROVES EFFECTIVE

THERE are numerous methods of preserving eggs, all of them of commercial importance, because, were it not for them, the market would not be relieved of its surplus at the time of the greatest production, and prices would fall so as to leave no profit for either producer or dealer. A novel industry is that of canning eggs, as practiced by the large packing houses. The shells are removed and their contents sold for baker's purposes. Another method of preserving eggs is by drying the whites and yolks, which are then sold in powdered form.

Ofttimes the knowledge of a coming scarcity of fresh laid eggs makes it desirable for the housewife to keep eggs for several weeks or even months before using. After repeated requests the Rhode Island Experimental Station undertook to determine which of the numerous simple methods could best be utilized to economically and effectively preserve the surplus of eggs, produced in the spring for a few months, so that they might be used to advantage in the fall and early winter, when eggs are scarce.

The oxygen in the air is the chief promoter of the chemical changes wrought by action of the germs, hence the exclusion of the air excludes both the germs, and their supporting element. For this reason the success of most methods of preserving depends upon the absence of air. Of these methods the following were deemed worthy of a trial:

1. Water glass (a silicate of soda).
2. Salt.
3. Slaked lime and salt brine.
4. Vaseline.
5. Dry wood ashes.
6. Gypsum.
7. Powdered sulphur.
8. Brimstone fumes and sulphur.
9. Permanganate of potash.
10. Salicylic acid.
11. Salt brine.

In each test fresh eggs were used, carefully gathered and cautiously handled. When a liquid preservative was used, the eggs were carefully washed before being subjected to the process. For the parallel tests twenty eggs were used, as uniform as possible in size, color of shell and age, laid by fowls of one breed, treated alike as to food, range, care and management. During the trials the jars containing the eggs remained undisturbed on the floor of a cellar closet where the temperature ranged from sixty-two to sixty-seven degrees F in summer.

RESULTS OF TESTS

WATER GLASS

Water glass, or soluable glass, was diluted with water in the proportion of one part water glass to nine parts boiled water. On May 18, twenty Leghorn eggs were carefully washed and placed in a stone jar. Over them was poured the ten per cent solution of water glass. In this preservative they were kept a little more than ten months, until April 4. Result: Good, 100 per cent; bad, 0 per cent.

The shells of the eggs were very clean, owing to the alkaline nature of the solution; the air cells were not enlarged. Examination showed the whites of the eggs to be clear, but not so limpid as those of fresh eggs. The eggs appeared normal in color and condition. They had kept well for ten months, and proved to be suitable for table use.

SALT

Fine table salt used. It was packed in the jar to the depth of two inches. Twenty eggs were packed in the jar, small ends down, not touching each other, and closely packed in salt. Result, April 4: Good, 0 per cent; bad, 100 per cent.

For preserving eggs for a few months, however, this method may be recommended. It is simple, cheap, and for short periods reasonably effective.

LIME WATER AND BRINE

One pound of quick-lime and one-half pound of table salt were thoroughly mixed with four quarts of boiled water. After slaking and settling, the clear solution was drawn off for use in the test. On May 18, twenty eggs were covered with the liquid. At the end of the period the result was: Good, 100 per cent; bad, 0 per cent.

The surface of the shells was clean and clear. The air cells were not increased in size. The whites and yolks were normal in appearance. The whites beat up nicely, but had a slightly saline taste. This old-fashioned method of preserving eggs was again proved effective.

FAILURE OF OTHER METHODS

In this test of long duration (ten months and seventeen days), the remaining methods proved total failures.

To determine more fully the value of water glass as a preservative, four more tests were made. These four lots were placed in the solution May 20, and were kept until the fourth of April following, when they were pronounced nearly as good for culinary purposes as fresh eggs. Further tests proved only more conclusively that this silicate of soda was nearly perfect as a preservative of eggs. From 120 eggs not one was bad.

WATER GLASS SOLUTION

The keeping of eggs in a ten per cent solution of water glass for a period of nearly a year resulted so successfully that it was decided to continue the experiments and ascertain to what extent the solution could be diluted and still remain an effective preservative. The result of this series of trials is here shown in tabular form.

TEST NUMBER	METHOD	Per Cent.	No. of Eggs.	PERIOD Months. Days.	GOOD Per Cent.	BAD Per Cent.
Twenty-five	Water Glass	10	30	11—19	100	0
Twenty-six	" "	5	30	11—19	100	0
Twenty-seven	" "	5	24	11— 6	100	0
Twenty-eight	" "	3	24	11— 6	0	100
Twenty-nine	" "	3	20	10—21	100	0
Thirty	" "	10	20	10—21	90	10
Thirty-one	" "	10	20	9—15	80	20
Thirty-two	" "	5	20	9—15	85	15
Thirty-three	" "	3	20	9—15	100	0
Thirty-four	Pure Water		17	9—15	0	100
Thirty-five	Water Glass	2	17	8— 8	100	0
Thirty-six	" "	4	15	8—24	100	0
Thirty-seven	" "	1	14	8—24	0	100
Thirty-eight	" "	3	15	8—24	100	0

The expense of the water glass at sixty cents per gallon would amount to about two-thirds of a cent per dozen eggs.

EXPERIMENTS—LAYERS AND EGGS

This does not include the expense of the jar or other receptacle, which may be of stoneware, glass or wood.

Note.—Those who visited the incubator exhibit at the Pan-American Exposition will remember that in a small booth at one end of the building a persuasive talker was earnestly declaiming the merits of his "new and sure recipe for preserving eggs."

72—AN EXCELLENT COLONY HOUSE

On a table he had a three-gallon stone jar filled with eggs which he claimed had been in the colorless solution for nearly a year. He had testimonials to show that some of the purchasers of his "marvelous discovery" had kept eggs for more than a year, at the end of which time they were as good as when laid. He did a rushing business selling formulas at "$1 each, reduced from $2." "Buy eggs when they are eight cents a dozen, and sell them for forty cents when eggs are scarce," was his inducement. His preservative was a ten per cent solution of water glass.

* * * * * * *

The Agricultural College, Guelph, Ontario, Canada, made the following experiments in preserving eggs:

Several methods of preserving eggs have been tested during the year. The eggs for this purpose were taken early in June, and were tested in December. Many of the same methods that proved fairly successful last year were again tried.

Method No. 1—A solution was used composed of one part water glass (sodium silicate) and five parts water that had been previously boiled. This was a very strong solution, and unless an egg was absolutely fresh, it would not sink in the solution.

The eggs from this solution were of fairly good flavor, and all were well preserved.

Method No. 2—This was similar to No. 1, except that eight parts of water were used instead of five parts. The eggs in this were nearly as good eggs as those from No. 1. This is a good preservative where it is desired to keep summer eggs for winter use.

Method No. 3—This was composed of ten parts of water to one part of water glass. There were no bad eggs in this solution, but the eggs were inferior in flavor and in poaching quality to those kept by methods Nos. 1 and No. 2.

Method No. 4—This consisted of the same solution as No. 2, but in place of allowing the eggs to remain in the liquid they were removed after having been in it for a week, except the last lot which was put in the solution. This lot was allowed to remain through the season.

(a) The eggs, after being in the solution for a week, were removed and placed in an ordinary egg case in the cellar. They were all good when tested, but had evaporated considerably and were lacking in flavor.

(b) These were the second lot of eggs to be placed in the liquid. They were handled similarly to those in (a), and were about equal quality.

(c) These eggs were allowed to remain in the liquid. They were well preserved, being good.

They were scarcely equal in quality to those from No. 2 but were superior to those from No. 3.

Method No. 5—A lime solution, made as follows:

Two pounds of fresh lime were slaked in a pail and a pint of salt was added thereto. After mixing, the contents of the pail were put into a tub containing four gallons of water. This was well stirred and left to settle. Then it was stirred thoroughly the second time and left to settle; after which the clear liquid was poured over the eggs, which had previously been placed in a crock or tub. Only the clear liquid was used.

These eggs were well preserved; but those from the bottom of the tub had a decidedly lime taste, and the yolk was somewhat hardened.

Mr. W. R. Graham, manager of poultry department; in reply to a question wrote:

"I find water glass (sodium silicate) is variable in strength, and also that the English water glass is much thicker and stronger than the American. We find that the American as we get it will sink in a preparation of one part water glass to

73—SUGGESTIONS FOR A GOOD DINNER

six parts of water, and in some cases to five parts of water, whereas with the English it will no sink in a solution of less than eleven to twelve parts of water.

"We find it costs us about one cent per dozen to put eggs in the English water glass, diluting one to twelve, and paying 80 cents per gallon for the liquid"

PRESERVATION OF EGGS IN COLD STORAGE

SEVEN HUNDRED AND TWENTY MILLION EGGS STORED IN CHICAGO ANNUALLY—
HOW THE EGGS ARE HANDLED WHILE THEY ARE IN COLD STORAGE—COST OF THE
STORAGE—POULTRYMEN CAN REALIZE AN INCREASED PROFIT BY STORING EGGS

FRED HAXTON

CHICAGO is the greatest egg center in the country. Not even New York can outclass the western metropolis in this respect. The cause of its pre-eminence as an egg market lies in the fact that the greatest egg producing states are in the middle west, and the country merchants and commission men, who buy most of the eggs, find it more convenient to send their poultry products to Chicago. Iowa supplies the market with more eggs than any other state in the union, its annual output being estimated at 100,000,000 dozen. Illinois stands third, with a supply of 86,000,000 dozen. In connection with this it is interesting to know that the United States produces 16,000,000,000 eggs each year, valued at $145,000,000. These figures are based on government reports.

On an average 2,000,000 cases of eggs come to Chicago every year. The cases hold thirty dozen eggs each, making a supply of 60,000,000 dozen, or 720,000,000 eggs each twelve-month. The arrivals fluctuate, decreasing materially as the season advances. During March, April and the early part of May about 20,000 cases are received daily, according to figures of the Chicago Butter and Egg Board. For the rest of the year an average of 5,000 cases a day is received.

Eggs are shipped to Chicago from all the states west of the city to the rocky mountains and south to and including Texas. Some states east of Chicago send their supply to its market, such including Kentucky, Tennessee and Alabama, but eastern states for the most part send their poultry products to New York and Boston.

SEASON FOR COLD STORAGE

The large poultrymen and the country and city buyers begin to send their eggs to the cold storage houses early in March. For three months the shipments are heavy. By the end of August practically all of the eggs have reached the coolers. The following month they begin to go, and the coolers are almost emptied by the latter part of January. Most of the cases go out in September, October and November. By the middle of December the bulk of them has been sold, for it is not considered safe from a financial standpoint to hold them later than that, though the men who had eggs in cold storage last January and February sold them at fancy prices. It will be remembered there was a scarcity of eggs at that time. Usually it is considered risky to carry the stock in storage after December 1.

The season in the western states opens first in Missouri, Illinois and Kansas. Supplies come next from Iowa, then from the Dakotas. Ohio sends most of its eggs east. In fact, the tendency in the states of the middle west is more and more to place eggs in storage in eastern houses, at least, as many as the dealers expect to sell there, as it is found there is less shrinkage—loss from broken and spoiled eggs—when the supply is shipped fresh. Eggs become weakened by several months' storage and are apt to be broken in transit.

Producers now are paying 40 cents a case for having their eggs kept in cold storage throughout the season, this being at the rate of little more than one cent a dozen, although one Chicago house makes it 30 cents a crate. The general rates for storage by the month are 10 cents for the first and 5 cents for each succeeding thirty days. These rates are well established, and represent a large reduction from those current a number of years ago.

When a producer ships a quantity of eggs he notifies the warehouse, which has the car switched to its plant on arrival in the city, and the eggs at once are transferred to the coolers. These are rooms usually thirty to forty feet wide and long. The cases are piled in tiers several inches from the wall, with slats between, to allow free circulation of air, and the tops of the crates are taken off. The main efforts are to keep the proper temperature and to maintain fresh air.

THE METHODS OF COOLING

There is little difference in the various systems of cooling the rooms, the ammonia device being generally used, although many of the newer houses are installing the carbolic acid cooling apparatus. Gases admitted into pipes reduce the temperature, but to describe the plan would involve a lengthy technical article. There is a great variety of opinion regarding the proper temperature. Some hold that with the circulation of cold air it is not necessary to keep the thermometer extremely low, and that with medium cooled atmosphere results are better. The time was when 40 degrees were considered low enough. Some coolers now are kept at the low point of 27 degrees. The average is

74—A LARGE EGG-CANDLING MACHINE

Examining Eggs with the aid of a special Egg-Candling Machine. Twenty-six thousand eggs an hour are tested. At one end of the covered booth is a shallow trough eighteen inches wide and three inches deep, in which the eggs are deposited. On the bottom of this trough there is an endless rubber belt which revolves and carries the eggs along the "chute" into the "inspection" booth, where they are deposited upon a set of twelve grooved rollers which are constantly revolving. Beneath these rollers there are four powerful electric lights with reflectors which throw the light upon the eggs, causing them to become transparent if good. Inside this booth are stationed four experts, whose duty it is to detect the defective eggs, also those that are cracked, as they pass along the rollers. The good, defective and cracked eggs are placed in different chutes and cared for by the packers. At the wheel on the upper right hand corner of the machine may be seen the "controller" who regulates the speed of the machine. When he finds the inspectors are getting behind because of the large number of defective and cracked eggs, he slows down the machine to enable them to catch up.

from 30 to 34. The white of an egg freezes at 25 degrees, and the safety point is placed at 27 or 28. If the eggs are stored in hot weather, the temperature is kept as low as possible without cracking the shell. If the eggs are to be cared for during only a short time, 40 degrees usually is considered sufficiently low. It is argued that a low temperature comes nearer preserving the egg in its normal condition; that it is unfavorable to evaporation and therefore does not shrink the

75—CASES OF EGGS IN COLD STORAGE

contents as much as the higher temperature. It would thus seem preferable from every point of view save that of expense.

The Chicago coolers have a capacity for millions of cases. More or less secrecy is maintained by cold storage houses as to the number of crates they have in keeping, and thus it is almost impossible to estimate accurately the number of dozen in storage at any particular date. One of the largest storage companies in Chicago stated it had received this season 240,000 cases. Last season it stored 300,000. The difference in figures is due largely to the fact that more eggs were shipped east at the beginning of the selling period than is usual.

The experiment of seeing how long an egg can be kept good apparently never has been made. Warehousemen say it has been demonstrated that deterioration can be arrested indefinitely, but that is the limit of their knowledge on this subject. It is not unusual, however, although not customary, to keep eggs in storage from the first of March until the first of the following February, and they are then in perfect condition. Much depends upon the condition of the egg when placed in the cooler. Take supplies freshly gathered in March, April or May, when they are best for cold storage purposes, and place them in the coolers at once, and there is no known reason why they should not be good at the end of two years, if they receive proper attention.

THE BUSINESS VALUE OF COLD STORAGE

The value of cold storage lies not alone in the fact that it preserves eggs during the season of plenty against the time when the hens are not producing so rapidly. It acts as a balance wheel on the market, preventing a glutting of the commission houses by too great shipments, and thus tends to the maintenance of uniform prices. In sustaining the quotations on eggs the year around it performs an invaluable service for the poultryman. In the days before it held its present recognized position in the field of produce, the farmer during the spring and summer months was obliged to sell his eggs for anything that was offered. He often was compelled to make great sacrifices in the matter of prices in order to induce the consumer to take them off his hands before they spoiled. Then when the cold months came and eggs were scarce, and prices absurdly high, the dealers had none to offer because there was no way of preserving them through the hot weather. Now the dealers are not forced to get rid of their supplies for what they will bring when the hens begin to lay in the spring. They simply release enough eggs to keep the market supplied, thus keeping prices up, and send the rest to the storage house. Then, when winter comes, the buyer has eggs enough to meet the consumers' demand.

Few of the egg producers handle their own stock directly in these days of combination and centralization. At various places in the great egg producing districts are located what are known as "central stations," where the commission merchants receive the supply from the farmers and storekeepers, candle and pack them, and send them to the storage houses, there to be held until sold.

When a buyer wishes to secure a lot of eggs, the commission merchant authorizes him to inspect the stock in storage to see if it is up to claims. The buyer takes an expert candler, gets down five or ten cases of eggs, and tests them. If the eggs have been stored early in the season the loss from bad ones and from shrinkage will be small.

Shrinkage means the diminution of the white of the egg. As the shells are porous, the longer the eggs lie in storage, the greater the evaporation of the albuminous substance. Especially is this true of the thin and watery eggs. The average shrinkage has been estimated at from three to six per cent. Eggs gathered in April, when stored in first-class coolers, suffer least from evaporation. When these eggs are kept where produced, the shrinkage may be reduced to three per cent.

PREPARING THE EGGS FOR STORAGE

So much of the success of the cold storage depends on the manner in which the eggs are packed, that the poultryman and shipper should exercise the greatest care when he sends his product direct to the coolers. The commission men at the centralizing points employ the best candlers and packers, and

76—CANDLING AND TESTING EGGS BEFORE PLACING IN COLD STORAGE

their methods of preparing eggs for storage are the best known, but the pultryman, in attempting to pack his own eggs, often makes blunders that leave him hundreds of dollars out of pocket.

We have heard of one farmer who not only packed his own eggs, but stored them in his own coolers as well, who lost hundreds of cases because he put the eggs in with the blunt end down. When he inspected them a few months later he found most of them bad. There is a scientific reason for this

EGGS AND EGG FARMS

result. As every one knows, there is a small hollow at the blunt end of the egg, between the shell and the skin that envelops the albumen and yolk. When you pack the eggs blunt end down and let them stand for months the weight of the meat is almost certain to break the thin skin and let the white and yolk drop into the hollow. Then decomposition begins. Pack them point down and this cannot happen.

TURNS CASES FREQUENTLY

One small storage man in writing of his experience said that he finds he gets best results by turning the cases upside down every few weeks, but most storage houses believe this is needless trouble if the eggs are packed point down.

Large, heavy shelled eggs should be picked for No. 1's, and they should be clean. The utmost care should be used to prevent a bad egg from finding its way into the case. One bad specimen can contaminate many good ones during a season's storage. On the bottom of the case a layer of excelsior or clean hay or straw should be laid. The eggs should then be packed in layers with hay, straw or paper between each layer. The interstices between the eggs may be filled with clean, dry chaff or the odorless pasteboard fillers manufactured for the purpose may be used. At the top another layer of hay or excelsior should cover the eggs, as at the bottom. Whitewood or paper boxes are considered the best for cases.

77—INTERIOR OF COLD STORAGE ROOM

Section of Egg Storage Room in Chicago Cold Storage showing central and rear ventilating air ducts. The cold air forced through pipes in the wooden shaft where the man is standing is let into the room through adjustable ventilators. There are three such air ducts in each room—one at center and one at each end.

THE COST OF CANDLING

It is difficult to compute the cost of candling eggs, according to "Ice and Refrigerator," of Chicago. It depends on the ability of the men doing the work. Some men and women candle with both hands at once, and, of course, are able to handle nearly twice as many eggs in a given time as those who are able to test but one egg at a time. A simple candling contrivance is made by boring a small hole in a box and placing a lamp, candle or electric light behind the opening. By holding the egg between the hole and your eye, you are able to determine whether it is good. It pays always, however, to secure the services of an expert candler if many eggs are to be put up.

THE COOLER MUST BE ODORLESS

The question whether or not other produce can be stored safely in the same room with eggs is a mooted one. Of course only such products as require the same temperature as eggs can be cared for. Storehouse men who are crowded for space often are tempted to place other stock in a partially filled egg cooler, but this is accompanied by danger. The best storehouses never try the experiment. They would prefer to turn the produce away. Certain it is that eggs stored near cabbages or onions will in a short time contract the flavor of these vegetables. One expert says that evaporated fruit will not flavor eggs, but others differ with him. The egg room should be as nearly odorless as possible, and the floors of nearly all these compartments in storage houses are liberally covered with lime to kill odors.

LIABILITY OF WAREHOUSEMEN

Much has been written in the cold storage papers concerning the liability of warehousemen for losses when for some reason or other large quantities of eggs have been found to be spoiled when taken from the coolers. Higher courts have ruled that the limited liability clause in the contract between the cooler man and the commission merchant does not remove the responsibility if the storage house openly is at fault. For instance, if the warehouseman agreed to keep the temperature in the egg room at or under a certain degree and through some fault of the machinery the temperature rose above that point and remained there long enough to cause deterioration to begin in the eggs, damages doubtless could be collected.

Cold storage opens a great field for the poultryman. Instead of selling his eggs at a low rate in the summer months, he can buy cases for around 25 cents each—less than one cent a dozen for the eggs stored—and place in the coolers, which the first buyers or the middlemen doubtless would do otherwise. All that is necessary is to write to any storage house in a nearby city and secure a blank contract giving the cost and terms under which eggs are preserved, and ship his product by express direct to the storage house. He can secure the name of a reliable firm by writing to any commission man in the city in which he intends to have the eggs placed, and when it comes time to sell the product he can drop a postal to the broker, ordering him to take out so many crates and sell them at the highest market.

OPENS POSSIBILITIES

The profit for the poultryman will be large, and the storage men will treat the small consignment with the same consideration that they give to the large. The average price of cold storage eggs is around two or three cents lower than that for the fresh article, although it is a fact that many stored eggs often are sold for the fresh and bring the same rates.

PROFITABLE EGG FARMS

RHODE ISLAND EGG FARMS

LITTLE COMPTON POULTRY PLANT AND COLONY PLAN DESCRIBED — PRACTICAL EGG FARMS FOUND IN LITTLE CORNER OF NEW ENGLAND WHERE THE POULTRY INDUSTRY HAS PAID GOOD DIVIDENDS FOR NEARLY TWO-THIRDS OF A CENTURY

PRINCE T. WOODS, M. D.

THROUGH some odd coincidence two of the great practical poultry growing sections in New England are both termed the South Shore. This fact has led to not a little confusion on the part of visitors and some of the natives who have written about these remarkable poultry districts. As a matter of fact the Boston market practically recognizes only one South Shore, the South Shore district which lies at the south of Massachusetts Bay and is made up of the towns of Hingham, Norwell, Rockland, Assinippi and adjacent nearby territory. This is the famous South Shore soft roaster district in which the majority of the toothsome fancy roasting chickens are produced that bring such long prices in Boston and Providence markets. This Massachusetts South Shore district is almost entirely devoted to the production of market poultry, or of eggs for hatching with which to supply the roaster growers.

Singularly enough the egg farming district of southeastern Rhode Island and the immediately adjoining portion of Bristol county, Massachusetts, is also known locally as the South Shore district or the South Shore of Rhode Island. This district is bounded on the west by the Sakonnet River, on the south by the Atlantic Ocean, and on the east and north by the Westport River and the Massachusetts line. Besides being termed the Rhode Island South Shore district this territory is often spoken of as the Little Compton poultry district, Little Compton being the chief poultry center, but the poultry territory also includes Tiverton Four Corners, Adamsville, The Commons, and South Shore, Rhode Island, and Westport and Central Village, Massachusetts. Here it is that the now famous Rhode Island Reds were originated and perfected, and it is from this section that they received their name. Poultry raising has been a prominent industry in this part of New England for nearly two-thirds of a century, gradually growing in strength until now it is probable that the chief farming industry in that little corner of New England, occupying a territory practically ten miles square on the shore of the Atlantic Ocean between the Sakonnet River and the mouth of Buzzard's Bay, is the colony farming of poultry for egg production.

Prominently mentioned among the pioneers in this business sixty or more years ago, we find the names of William Tripp and Isaac C. Wilbour of Little Compton, Rhode Island, and John Macomber of Westport, Massachusetts. The late Mr. Tripp is credited with being the originator of the Rhode Island Red fowl, and these birds which were intermixed with red fowls produced by Mr. Macomber were originally known to the residents of that section as the "Bill Tripp" and "John Macomber" fowls. So far as can be learned these ancestors of our present day Rhode Island Reds were the result of crossing Malays with birds that were then termed Chittagongs, Cochin Chinas and Red Shanghais. Messrs. Tripp and Macomber both took a great interest in developing these red fowls chiefly for egg production, but with a view to producing fowls that could be disposed of advantageously as market poultry when through laying, and for this reason clean legged cockerels and pullets were chosen with bright yellow legs and yellow skin, and also the breeders were selected from those pullets which produced the greatest number of large brown eggs. As interest increased in these fowls care was also given in the selection of breeding birds to secure those most even in plumage, of good shape, and not too coarse boned.

RHODE ISLAND REDS

Mr. Isaac C. Wilbour was one of the veteran poultry raisers and handlers of the Little Compton district, and he with Dr. N. B. Aldrich of Fall river was instrumental in introducing the Rhode Island Reds to lovers of fancy poultry, in fact Dr. Aldrich is credited with first applying to these birds the name Rhode Island Red. Up to that time they had been spoken of chiefly as "Red Malays," "Reds" and "Macomber" or "Tripp" fowls. It speaks well for the merits of this red bird that today the visitor will find very few other fowls in the whole Little Compton or Rhode Island South Shore district.

Just before Christmas, 1906, we had the pleasure of visiting this interesting egg farming district, and drove for nearly 35 miles over rough country roads endeavoring to see as much as possible of the colony house egg farm plan of poultry keeping as we could in the brief amount of time at our disposal.

EGGS AND EGG FARMS

Although the weather was not agreeable and not in the least favorable to viewing country plants or taking photographs, we enjoyed the trip so much, and took such great interest in what we saw, that we made plans before leaving for home to make another visit to this great egg producing center under more favorable conditions in the spring of the year when the work of growing chicks to produce next season's layers and breeders will be in full swing.

A VISIT TO LITTLE COMPTON

Leaving the South Station in Boston at 6:35 in the morning, we arrived at Tiverton, Rhode Island, which lies at the head waters of the Sakonnet River, near the point where they join Mt. Hope Bay, shortly after half past eight and here secured a livery outfit to drive through the Little Compton district, for one must go several miles from Tiverton Station before opening up the poultry growing country, and the largest egg farms of all are located ten miles or more from the nearest railroad station. To the visitor who is unused to country roads in winter, we recommend visiting this section in the season when the green is in the grass and on the trees and

of some of the other poultry raisers in his section. Nearly all of the eggs handled by him go to the Boston market to special high class dealers who handle only first quality high priced eggs and provisions. This is still quite an extensive business in the Little Compton section, although of late years individual egg farmers are to a greater extent finding their own special makets for their output instead of depending upon a local handler.

After leaving Tiverton Station we drove along the shore of the Sakonnet River without seeing much of interest in the poultry plant line until we began to approach Tiverton Four Corners. From here on, on either side of the road, we saw farms large and small with anywhere from 10 to 50 small colony poultry houses scattered through the fields. As we approached Little Compton it seemed as if every farmer in the entire district must be devoting the greater part of his energies to keeping poultry, in fact, the greater part of the district makes one think of a great poultry city, so thickly scattered are the poultry buildings.

Having met Mr. Fred Almy of Little Compton a few days

78—VIEWS OF TYPICAL RHODE ISLAND COLONY HOUSES

the roads are smooth and fine. In the summer season better roads than they have in this district will not be found in any farming section, but the winter road service is none too smooth. A stage line plies daily between Tiverton Station and Little Compton, and the entire district is covered by an excellent rural free delivery mail service. Nevertheless it would be an eye opener to some of those poultry theorists, who believe that to be successful it is essential for a poultry plant to be near a railroad center or central shipping point to visit these large poultry ranches scattered over the country far from a railroad station and remote from the base of supplies. The railroad stations for this district are Tiverton, Rhode Island, and Westport, Massachusetts. In addition to these points from which eggs may be shipped and grain received, a steamer leaves every other day from Sakonnet Point for Providence, Rhode Island, a salt water trip of nearly 40 miles. Not the least among the cargo of this little steamer are the many cases of high grade fresh eggs which are shipped to supply the markets of the city of Providence. Boston, Massachusetts, also receives some of the output of this section.

Mr. P. H. Wilbour of Little Compton, son of the late Isaac C. Wilbour, continues the business developed by his father, operating a poultry plant or egg farm of his own in connection with the business of handling the egg shipments

prior to our visit, at a Convention of the Connecticut Board of Agriculture, we made his plant our objective point and reached his place of business shortly before noon. Mr. Almy was on hand to welcome us and never have we had a more enjoyable visit at a poultry plant, even though we had to make it an exceedingly brief one because of our desire to see more of that poultry section.

LARGE PRACTICAL EGG FARM

Mr. Almy operates a farm that contains 125 or more acres and so far as we could see it is almost exclusively devoted to poultry. Here, although the weather was cloudy and threatening rain and snow, we made a number of snapshots of poultry buildings that are reproduced with this article. These show the typical Rhode Island colony houses for layers and colony brooder houses or coops for hens with chicks. We wish it were possible to have some of the many theorists and would-be scientific poultry keepers visit Little Compton and spend a week absorbing the sound common-sense practical poultry keeping which is in vogue there. Such a course might perhaps be considered radical treatment and might result in exploding some theories and disproving many presumably "accepted scientific facts." Certain it is that the methods employed here are much the same as they have been for near-

PROFITABLE EGG FARMS

ly three quarters of a century. They are plain, simple methods, somewhat old fashioned, almost entirely free from theory and intensely practical.

In Little Compton district poultry farming has been an established business for two generations and more, and they make the business pay. One does not need to ask in this section "Is the business profitable, will poultry pay?" when on every hand he sees evidence of prosperity among men who are poultry farmers and where nearly every farm carries from several hundred to two or three thousand head of laying stock. Within itself this is sufficient evidence that poultry keeping pays and pays well, even when run exclusively for egg production, where the fowls are kept according to the methods employed on these plants.

Mr. Almy has been in the business for about fifteen years and is reckoned as one of the most successful young men in the neighborhood. Possibly we have a prejudice in favor of tall men from a common bond of sympathy between us, but be that as it may, we took a liking to Mr. Almy at first sight and liked him better as we became more acquainted. As we spent the most time on his plant we will endeavor herewith to give a brief outline of his methods.

PRACTICAL COLONY HOUSES

To begin with, the Almy plant carries on an average two thousand head of breeding and laying stock, all with the exception of half a dozen pens being kept on the colony plan. It is probable that all the birds would be in colony houses except for the fact that there was one long low building on the place when the proprietor made his start. This building was converted into a breeding house and has since been occupied by selected breeding stock, the only penned and yarded fowls on the farm. There are some 50 or more colony houses on this plant of varying styles and shapes, the prevailing pattern being a double pitch roofed colony house 8 by 12 feet in size, 6-foot post or stud at the eaves, the double pitch roof having a sharp incline bringing the peak about 8½ or 9 feet from the ground. These poultry buildings are all economically built, it being the belief of most of the poultry growers in this section that the initial cost of a poultry building should not be more than 50 to 65 cents per bird, that is, a poultry building costing from $20 to $25 should house 40 fowls, and such a building should be so constructed that it ought to last from 20 to 25 years. This reduces the cost of poultry houses to the minimum and saves involving a considerable amount of capital in buildings which might otherwise seriously handicap the poultry venture at the start.

All of these colony houses are built on runners, that is, the long sills are made sufficiently deep and heavy to stand having eye-bolts and chains fastened to them so that the houses can be dragged about by horse power from one part of the farm to another, as may be desired. This easy moving feature of these colony houses is a valuable one, as it permits changing the houses about from field to field and growing

79—RHODE ISLAND COLONY BROOD COOPS ON FREE RANGE

some farm crop on the vacated range season by season, thus keeping the soil sweet. Although as a general rule this moving is not practiced as often as it might be expected, some of the buildings now occupying sites close by where they were originally erected. Owing to the rolling character of most of the land, it is easily drained and does not become seriously fouled through constant use, since the wash of the rains cleanses it. On flat fields or ill drained land these portable houses become a valuable factor in insuring the success of the plant and the greatest production of sturdy fowls and good field crops. Being built with a view to moving about in this manner, all houses are stoutly framed and on this frame are used common planed inch boards. Some use matched boards, others use common sheathing boards and batten the cracks, and still others do not batten the cracks at all, leaving them open. Mr. Almy shingles the roofs of all his buildings, or covers the roofs with roofing paper, but many farmers in this section make the roofs simply of boards and battens or use matched boards without battens. The tight roof is about the only tight thing about the colony houses as used in this section. Windows are so placed that they can be readily removed or opened. The roosts run the short way of the house and are located in the end away from the door. In the majority of the houses this brings the roosts in the west end, although many of the buildings have the roosts in the east end and door in the west front. The fronts of the houses are on the 12-foot south side and contain the windows and door, two 12-light sash being the usual accompaniment for an 8 by 12 building having a double pitch roof.

As we said before, the cracks in the sides of the building are not battened, as a rule, and to avoid drafts about the roosts a strip of building paper is tacked on to the inside of the house with lath battens immediately around the roosting quarters. Only one strip about a yard wide is used. Some of the buildings do not even have this protection about the roosts and there are cracks through which one can easily see outdoors. Each house is provided with a small door for the fowls on the floor level, which is never closed. In addition to this there is the regular sized door of wood and a screen or slat door by which the fowls may be confined when the solid door is open. These are all plainly shown in the illustrations. In addition to this the double pitch houses have a small ventilating scuttle or door in the north wall just below the eaves. This is used only in summer time for purposes of better ventilation.

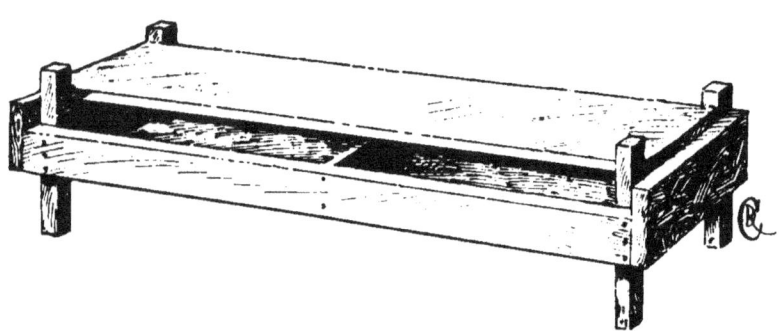

80—RHODE ISLAND COLONY FOOD TROUGH

From 40 to 50 breeders or layers are kept in one of these houses. No dropping boards are used. The roosts are placed at convenient height from the floor, the floor is of earth built up of sods to the level of the sills, and then filled in with three or four inches with salt water beach sand from the shores of the Sakonnet River. This sand takes the place of litter and is used chiefly as an aid to keep the floor clean so that the birds will not get their feet dirty and soil the eggs.

LAYERS HAVE FREE RANGE

While these buildings are made so that they can be readily moved about to any part of the farm, they are seldom moved but are cleaned out regularly, spring and fall, and as the proprietor told us, oftener if they need it, but it is seldom necessary to clean out the buildings more frequently than twice a year. Cleaning one of these houses is an easy matter, as all that is necessary is to elevate one end of the long sills a little, attach a chain and with a farm horse drag the house to fresh ground. The floor is then filled in with fresh sods and earth to the level of the tops of the sills, and on top of this is spread a few inches of beach sand. The accumulations of droppings and fouled soil on the site previously occupied by the house may then easily be loaded on a wagon and carted to any part of the farm where it will do the most good. Sometimes the buildings are cleaned without moving them.

The fowls run at large at all seasons of the year having absolutely free range except when they keep to the house of their own accord when the ground is covered with a heavy fall of snow. The colony buildings are located anywhere from 25 to 50 or 100 feet from one another, and there seems to be no difficulty about the flocks mixing to any great extent. The pullets or hens invariably return to their own houses to lay and the male birds, there being two or three to every house, seldom go far away from home. It is a peculiar fact that the males are not particularly quarrelsome. So long as the male bird keeps by his own homestead he seldom gets into trouble, but if he visits a neighboring house he is usually so badly whipped for his pains that he seldom repeats the offense.

The eggs from fowls kept on this plan show remarkably good fertility and at the time of our visit the birds were all in the best of health and condition, in fact we have never seen a flock that as a whole were any better as to health and condition than were these free range Rhode Island Reds.

ANOTHER TYPE OF COLONY HOUSE

In the operation of all these colony houses the large board door is opened early in the morning and left open until after the eggs are collected at night, except in extremely cold zero weather when the eggs would be likely to freeze if the door were left open. The small poultry door is never closed. When the fowls are first placed in these buildings they are confined by small wire runs that are quickly strung up on temporary posts driven into the ground in front of the building. After a few days' confinement close to the house they soon become wonted to their quarters and seldom occasion any trouble by mixing up with the other flocks, always returning to the home houses at roosting time.

Many of the farmers in the Little Compton district feed according to the dry food hopper plan, but the Almy plant is a notable exception. Here the birds receive a moist mash in addition to dry grain, but they are fed only once a day. The mash is mixed in a large cast iron cook kettle that will hold about 150 gallons. In this kettle the cut clover or vegetables which are to be used in the mash are scalded or cooked. Into this liquor the ground grain is stirred with a long handled post-hole spoon, the wooden handle of which is armored with iron to prevent wear.

HOW LAYERS ARE FED

The mash mixture is usually made up as follows: Bran and oat feed, equal parts, with sufficient coarse corn meal or fine cracked corn to make the mash dry and crumbly. About 20 per cent beef scrap and from 40 to 50 per cent clover or vegetables is used in this mash in winter time. In summer time fresh refuse fish from the fisheries in Tiverton is used to quite a considerable extent to take the place of beef scrap, and it is believed that the fish makes a very desirable meat food for both growing chicks and laying fowls. Care is taken not to feed a sufficient amount to "taste the eggs." This same mash which is fed to the laying fowls and breeders is also fed to growing chicks after they are two or three weeks old. In addition to the mash the birds receive a dry grain mixture of three parts corn, one part oats, wheat and barley, the quantity of the last three grains being varied from time to time according to convenience and the supply. Feeding and watering is done once a day.

All food is carted about the plant by horse power on a low flat wagon similar to a stone bogie. The mash is carried in barrels on the rear part of the wagon and the dry grain in bags or barrels according to convenience. The water is also taken to the birds on the same trip and this is carried in 8-quart milk cans tightly stoppered and placed in wooden racks to prevent upsetting.

PROFITABLE EGG FARMS

Immediately after breakfast in the morning they start out to feed the breeding and laying stock. In each house is a wooden trough on short stout legs with a board cover to prevent the birds from jumping on top of the food, a sufficient amount of space being left between the sides of the trough and the wooden cover to allow the birds to have free access to the food. These food troughs vary in size but on an average are three and one-half feet long by about 14 inches. A sufficient amount of mash and of dry grain food is left in each one of these troughs to supply each flock for the entire day. Experience has taught so well the requisite amount that it is very seldom that there is any considerable amount of grain left over, but in feeding care is always taken to see that the birds have a plentiful supply.

Mr. Almy prefers this plan to hopper feeding because there is not any considerable amount of food about the poultry house at nights to encourage rats harboring there, and in this section rats are the greatest pest and cause the most trouble. Where hoppers are used rats frequently take up their abode immediately under the poultry house, and are very difficult to get rid of, and these vermin do most of their work at night.

At feeding time the large doors of the houses are opened and fastened back (in warm weather they remain open night and day, the slat door only being closed at night), and if the weather is mild the ventilating windows or doors may also be opened. The birds are not visited again until it is time to collect the eggs in the afternoon, when the wagon again makes its round, the eggs are gathered and the doors of the poultry buildings closed.

Early hatched Rhode Island Red pullets are the main dependence for egg production and birds are only very seldom kept longer than two winters. It is claimed for these practical utility birds that the average pullet will weigh not more than five pounds at laying maturity, and that as a yearling hen her weight will be about the same. The fowls culled from laying houses are chiefly disposed of alive as market poultry. Surplus cockerels usually find their way to market as broilers or as small roasters, but these are merely side lines of the main egg business.

VALUABLE POINTERS IN FEEDING FREE RANGE FLOCKS

The growing chicks on the Almy plant are all reared in large stone-fenced fields, some considerable distance from the farm buildings. All food and water must be carted to them. As these chicks all have free range and roam at will, it is necessary to adopt some plan to avoid their crowding at meal times. Right here Mr. Almy gives a valuable pointer to poultrymen whose birds crowd and cause trouble at feeding time. He contends and with good reason that when chicks or fowls expect to have the food arrive at and from a certain point each day, they are bound to congregate there in an endeavor to anticipate the coming of the food. That is where trouble starts from free range birds crowding together at feeding time. To avoid this he makes it a point to enter his chicken fields at a different point, as far as possible, each day when approaching his flock with the feed wagon. The result is that birds never know in which direction or from which point of the large range the food supply is likely to arrive, and they know from habit and experience that the only safe and sure way to get a full meal is to hang around home until it arrives at the regular feeding place for their particular flock. This plan has been carried out now for several seasons and proves entirely practical, so much so that crowding at feeding time does not cause trouble on the Almy plant.

Mr. Almy is a very conservative man and in conversation with us made it apparent that it was his desire to understate rather than to overstate the facts concerning the poultry business. When pressed as to the profit that could reasonably be expected per hen when kept exclusively for egg production, he said that the net profit would range from about seventy-five cents to one dollar per hen per year, and that he estimated that he was able to make on an average 75 cents clear profit per hen per year.

Where birds are kept on the colony plan as they are in this section, feeding and watering but once a day, one man and a boy can comfortably care for from 2,000 to 2,500 head of laying and breeding stock and raise from 5,000 to 6,000 chicks in a season, and still have time for other necessary farm work. On all of these Little Compton plants the greatest economy is practiced in labor and in housing, two very important items of expense on any poultry plant.

At the time of our visit in the latter part of December the grain supplies for the laying and breeding stock on this plant were averaging to cost between $40 and $45 a week, of itself a considerable item but by no means as great as the cost of food when the chick season is on, at which time frequently five barrels of mash food besides a large amount of hard grain are daily required to supply young and old stock. This food cost,

81—ROW OF COLONY HOUSES

however, does not appear so alarming when one considers that December eggs in that section averaged 45 cents per dozen at wholesale; and the daily egg yield for the plant averaged about 50 dozen per day or a thirty per cent yield for the entire flock.

Our drive from the Almy place covered a wide stretch of beautiful rolling open country plentifully dotted over with colony poultry buildings. To give the names of all those who raise poultry for market eggs in this section would make this read like a directory of the Little Compton district. Nearly every farmer carries several hundred laying fowls any many of the farmers carry from 500 to 3,000 or more head of laying stock. Among the prominent poultrymen of today as in the earlier days of the industry will be found the names of Wilbour, Sission, Howard, Simmons, Briggs and Almy.

There is much to see and much to learn in this little Compton or Rhode Island South Shore district, and any persons who have a sufficient interest in poultry and the means and time at their disposal will find a visit to this section will amply repay them. There are none of the frills and furbelows that sometimes attach to recently developed hen fever, but for rugged simplicity with good results in practical poultry keeping, not to speak of theory exploding also, this small seagirt corner of New England has few if any equals in the great poultry industry of the United States.

POULTRY AND EGGS FOR MARKET

AN ARTICLE CONVEYING AS MUCH VALUABLE INFORMATION AS CAN WELL BE GIVEN IN THE SPACE OCCUPIED—HOUSING AND FEEDING—ONE HUNDRED DOLLARS FROM A BERRY PATCH AND AS MUCH MORE ADDED TO THE VALUE OF THE FOWLS

H. J. BLANCHARD

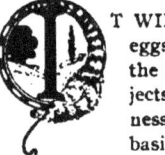

IT WILL certainly pay well to raise poultry for eggs and market only and pay no attention to the fancy, as eggs and meat are the prime objects of poultry keeping. My own poultry business was started and well under way on this basis alone. In most markets fine, fresh eggs pay much better than broilers or roasters, and the money comes in steadily the greater part of the year.

It is well to start with only what fowls can be properly housed and cared for, and increase the number as experience and judgment prompt. A few fowls well kept will pay better than many when crowded and neglected. The ideal way to keep a large number of fowls is to treat each flock as if it were the only flock you possess.

Eggs can be produced at a good profit in spring and summer because even though the price is then low, the cost of production corresponds. Winter egg production is highly profitable if properly managed. The requisites are warm, dry, and ventilated houses, well selected food, including succulent green stuff, and judicious exercise. Fancy new laid eggs usually sell best in a large city, and it would be well to locate within twelve hours' shipping distance of the market. Such a distance from the city would enable one to secure a suitable piece of land cheaply. The soil should be well drained naturally. Rough land will do just as well for the range, although some of the place should be good, tillable soil to enable the poultryman to raise a part at least of the food for the birds, and so reduce the regular expense for food stuff.

82—AN INCUBATOR HOUSE (ENGLAND)

For best results I favor the colony house, free range plan. Be sure to have houses far enough apart so that each flock gets plenty of range. This will make you more work in caring for the poultry, but you will be well paid for it. Feed just as well as you would if the birds were yarded and you will get better results.

FREE RANGE ADVOCATED

In my opinion, the best grain to grow on a poultry farm is corn. Plow under a liberal coating of manure from the hen houses or stable and give the crop thorough cultivation while it is young. Set your outdoor brooders near the cornfield and let the young chicks have free range through it. This will afford them shade, and protection from hawks and crows, besides being a grand foraging ground for them. The birds also help to keep up the fertility of the soil. We think corn one of the very best foods for poultry, especially when fed in connection with wheat. A blackberry patch is also a grand place for chickens to run in. It affords shade and protection all the season and in it the birds find a great many bugs and worms. The poultry also help to keep down the weeds. They seldom eat many of the berries as they grow too high. All of our cull hens as well as one line of our breeders are yarded, and in most of the yards we have Snyder blackberries growing. How the hens do enjoy wallowing in the shade of these bushes, and what fine berries we get! Our half-acre blackberry patch this year netted us over $100 and I believe was worth nearly as much to the chicks. Another very important crop is mangel wurzels. We formerly used cabbages for winter green food but found them very difficult to keep late in the winter. Select a smooth, fairly level piece of ground, free from large stones, and early in the spring plow under a heavy coating of the cleanings from the hen houses. Harrow thoroughly until fine and mellow. Mark rows three feet apart, using a light marker. Then scatter a liberal amount of commercial fertilizer in the mark and mix with the soil by dragging a stick back and forth. Then sow the seed liberally and evenly as possible by hand in the mark left by the stick and cover about an inch with hoe or rake. We formerly used a seed drill for sowing our mangels, but found the hand method far more reliable.

MANGELS FOR GREEN FOOD

We feed these beets to our poultry by simply cutting in halves or large pieces with a shovel and placing on the floor. The birds love the sweet, juicy roots and will work at them until all is eaten but the skin and frequently they consume that. The mangels should be harvested early in October before heavy freezing by simply lifting from the ground with the hands and breaking (not cutting) off the top, drawing to a frost-proof, well ventilated cellar, and piling in a corner or bin. We now have a pile seven feet deep in one of our cellars and they will keep fresh, crisp and juicy until next summer.

Our breeding stock and the layers are all fed alike, corn, oats, buckwheat, and wheat, even parts, being the morning and night food, with mash at noon. After over twenty years' experience in breeding and feeding White Leghorns we still believe in feeding them three times a day, with the mash at noon. White Leghorns are very active, and when fed judiciously three feeds a day are better than two. In the morning we give a very light ration of the mixed grains scattered in the litter on the floors. This keeps them busy for some time and gives them exercise. Next comes green food of some kind. Pure water is kept before them all day, and in cold weather the water is slightly warmed. Crushed oyster shell and grit in suitable places are accessible continually. At noon comes the mash, the most important feed of the day. The hens have been busy nearly all the forenoon scratching for grain in the litter, picking at mangel beets or cabbages and taking an occasional drink from the water pan, and are happy and hungry. In summer time the mash is moistened with skim milk, buttermilk, etc , and in winter with hot water. The mash is made of ground corn, oats, peas and barley, mixed with an equal bulk of wheat bran, and to this is added about one and

83—HOUSE AND SHELTER FOR GROWING CHICKS (ENGLAND)

one-half pounds of old-process oilmeal and three pounds of prepared meat for each hundred hens. The whole is then mixed thoroughly while dry. No condiments are fed except a little salt dissolved and added to the mash. The whole is then moistened and fed crumbly in troughs.

It is interesting to see the hens watch the doors and gates about feeding time, and when the mash is carried in they will fly into the pail and even onto the one carrying it. We feed all they will clean up and the amount they will eat varies greatly. Close observation enables us to determine about how much to feed. When the hens are laying heavily they usually eat a great deal more than when molting. Toward night they are fed all the mixed grain they can eat, scattered in the straw, and take more exercise scratching for it. If the birds do not act very hungry their morning food is decreased until they show a proper appetite.

STRUCTURE OF HOUSES

The houses are made with straw loft, and are warm and dry in winter, being well ventilated by sliding windows on every possible occasion. In summer the windows are all open and slatted doors used to give all the ventilation possible. Cleanliness, ventilation and kerosene keep these houses entirely free from lice. Some of the houses have a three-foot basement with ground floor, which is much appreciated by the hens, especially in hot weather. In winter each flock has a large box of fine road dust for bathing, which is often removed. The sanitation employed and the conditions surrounding these fowls, together with the method of feeding, render them very strong and vigorous. This, coupled with our method of selection in breeding, gives us typical laying stock. They have vigor and stamina to resist the usual diseases of poultry, and a good constitution to transmit to their offspring.

Our incubators are started about the middle of March and brooder houses and brooders are put in shape for the chicks. When a hatch is off they are put in a warm brooder so arranged that they can get just the amount of heat they need. The first few days they are fed oatmeal and grit with water slightly warmed and are kept in the brooder. In a few days they are given a room 16 by 20 to run in, the floor being covered with chaff or other litter. The chicks are gradually led on to a diet of johnnycake made of ground corn, oats and wheat, oat hulls sifted out. The chicks are soon given a little whole wheat scattered in the litter on the floor of the brooder house. Soon as the weather permits the chicks are let out into the yards, which are long and grassy. After the chicks are done with the brooders and have learned to roost, they are given free range, and the pullets separated from cockerels. The chicks are now fed a mash with a very little meat in it and the quantity of meat is gradually increased until the pullets begin laying, at about five months of age. During the fall the old stock is gradually sold for breeders and layers, and then the earliest pullets are put into the laying yards to take their places and supply our city market with eggs. We ship our eggs three times a week in summer and twice in winter, guaranteeing every egg to be new-laid.

The poultryman is well equipped for keeping up soil fertility on his farm. The chicks and fowls on free range contribute greatly to this end, and the manure from the roosts, houses, brooders and coops if properly applied will keep his farm very fertile.

84—RANGE OF SCRATCHING SHEDS (ENGLAND)

Hen manure should never be stored, as it is bound to lose a large amount of the free nitrogen, besides being a difficult, laborious job. We always draw it direct to the fields and spread thinly on meadow or pasture, according to season, or spread on the land intended for corn or root crops, to be turned under in spring by the plow.

A LARGE EGG FARM

PURE-BREDS BY THE THOUSANDS—THE FANCY AND MARKET BUSINESS HAPPILY COMBINED—OPEN HOUSES USED IN SUMMER AND IN WINTER—FEEDING FOR EGGS AND MARKETING—HOW IT IS DONE AT THE WHITE LEGHORN POULTRY YARDS

L. A. PECK

SEVEN THOUSAND hens on one farm makes it probably the largest egg farm in the world, and that it is purely and solely an egg farm is proved by the fact that no chickens are raised there; not any of the seven thousand layers being raised by Mr. Hayward himself—all are bought each year. It certainly is a remarkable story, and different from anything else, different from the story of every other poultry farm of which I have any knowledge,—and I have visited a great many of them. This remarkable egg farm is owned and operated by Mr. C. E. L. Hayward and is located in a small town of New Hampshire, about sixty-two miles from Boston. There is a good deal of up and down hill to many New Hampshire towns, and the road to Mr. Hayward's farm has so much of the up and down to it that heavy loads must be difficult to draw and the three miles' haul over such a road must considerably increase the cost of the about a ton of grain per day fed to the flocks. I visited the farm in winter some ten years ago, and Mr. Hayward has not only more than doubled his poultry business, but has made one decided change in method. At that time the farm was run on the free-range plan, the birds being given free range after a few days' confinement had wonted them to their quarters; now they are kept in close confinement, the birds never going out of their houses (after being put in in the autumn) until they are sold off to market to make room for the succeeding flocks. One reservation should be permitted in this general statement. The brood-coops (for breaking up the broodies) sit on the ground outside and in front of each house and persistently broody birds are put out in these coops for a few days until broken of the desire to incubate, then returned to the house again; by so much they are not closely confined to the houses, but to all intents and purposes the general statement is correct.

OPEN HOUSES PROVIDED

Another decidedly remarkable thing about Mr. Hayward's methods is that the houses are quite open to the weather, and are open just the same summer and winter. They have lots of snow in New Hampshire in winter; I have seen three feet depth of it! They sometimes have very cold weather in New Hampshire. I have seen it twenty-seven degree below zero there, and yet—there is a great poultry farm, with the fowls kept for egg production, the fowls being housed in small flocks in houses the fronts of which are half open and the birds exposed to the rigors of a New Hampshire winter; we will describe the method first, however, and discuss it afterwards.

The houses are alike as to plan, being eight feet square on the ground and built exactly like the "A" tents in which some of us slept in 1861-65. The floor is of two thicknesses of boards laid so as to break joints, and there is no frame whatever. There is a square base some fifteen inches high made of two inch thick planks, then the roof boards, cut eight feet long and nailed to the base and the inch-board ridge pole. The back (north) end is boarded up solid, while the front end is boarded up about fifteen to eighteen inches and down from the apex of the roof about eighteen inches (to give sufficient stiffness), and the balance is just sufficient boarding to make a door with a frame to hang and hasp it to; all the open space is covered with inch-mesh wire netting, which effectually excludes "varmints" but freely admits the air. The earlier houses had board roofs battened, then a hundred or two were built with roofs and backs covered with corrugated iron or some of the special roofing papers, but the latter were not satisfactorily durable and the corrugated iron drew the sun and made the houses uncomfortably hot during the summer. In recent years the houses are being built with roofs and back walls shingled, and as the earlier roofs need repairs they are being shingled also, so that shingled roofs and back walls will soon be the rule; in the country where hemlock, spruce and second growth pine shingles are so cheap as they are in New Hampshire it is almost surprising that shingles were not adopted sooner.

The houses rest upon four small stones, one at each corner, to bring the floors up from the damp ground; in that country there is much frost, and the freezing-thawing of the ground causes these foundation stones to sink into the ground gradually. We found two of the men busy prying up some of the houses and adding brick to the foundations, to lift them up again and prevent the floors from rotting.

INSIDE EQUIPMENT OF HOUSES

The inside furniture of these houses is of the simplest. At the back and about three feet above the floor two roost poles are set, about a foot space between them and the rear one about six inches from the wall. There are two small nest boxes, one in each front corner; a small box (about ten inches square by six inches deep) for the food, another for crushed oyster shells and a dust box about two feet square by eight inches deep. The water pan is outside, at the back corner, with a small gutter to convey the drip from one eave to it in rainy weather, and an aperture 2 by 4 inches gives the fowls room to put the head out and drink. The pan is of cast iron, is about ten inches square by five inches deep and is emptied and carefully rinsed out once a week. This arrangement of water pan and gutter entirely avoids watering in rainy weather; and in winter, when there is snow on the ground, a shovelful of snow is put in each house for the fowls to eat. In answer to our exclamation of surprise, Mr. Hayward said the snow did not seem to hurt them any, that it was the simplest and easiest way to water (?) them, and that they would eat snow if running out of doors—this last point will be readily understood by those who have observed fowls when they get access to snow. The broody coops were about two feet square, with a board floor and roof and slat sides. A small pan of water and a dish of food is set on a board in front of each coop, and the prisoners reach their heads through the slats to eat and drink.

Before the roofs and back walls of the houses were shingled we estimated the cost at eight dollars each; the present houses will cost about ten dollars each. At the cost of lumber at the present time this may be a bit too low, but the poorer grades of lumber are cheap in New Hampshire. These houses are set about two rods apart and (where the ground favors) in rows, which are about four rods apart. Sometimes stones or trees interfere with exact distances, but these are approximately correct. With twelve birds in a house it takes nearly six hundred houses for the seven thousand birds. A pathway for wheelbarrow runs alongside each long row of houses and a road for a horse and wagon along the ends of the rows. The food is put upon a cart and hauled along the roadway, a stop at the end of each row being made to load the wheelbarrow for one or two rows; there are eight to ten men employed, and each man has his group of houses to feed and water and take care of.

HOW FOWLS ARE FED

The fowls are fed twice a day, the morning feed being a mash, of which enough is fed to last for the birds to pick at till noon; the afternoon feed is usually wheat and occasionally corn. At the time of my visit they were feeding scorched (or damaged) wheat, from a burned elevator. This scorched wheat cost only $15 a ton on the track, and Mr. Hayward said the fowls ate it well and he could not see but that they laid as well as if fed the best of wheat. The morning mash is made up of a mixture of five hundred pounds mixed feed (wheat middlings), two hundred pounds corn meal, two hundred pounds beef scraps, one hundred pounds meat meal (making half a ton), two bushels cut clover, four quarts salt. The meals, scrap, etc., are all mixed together and got ready the night before, the salt dissolved in water and cut clover put to soak in water; before feeding the mash is made up by mixing with water, warm in winter and cold in summer, until the mess is moistened and is "crumbly" but not wet and sloppy, and is not "cooked" nor even scalded. It is upon the meat element in the above mixture that Mr. Hayward relies for his eggs, his expression being, "It's the meat that makes the eggs." That this is not the most completely "balanced" ration will be conceded, but its results are satisfactory to Mr. Hayward. the fact that he is still enlarging his plant and increasing his business is proof of that.

Mr. Hayward does not raise his laying stock. He has the pullets hatched and raised for him by arrangement with farmers in Vermont and New Hampshire. We recalled having seen his advertisement of "pullets wanted" a few years ago and he said he did not have to advertise now, that he could arrange beforehand and for all that he would need. He sells off the old birds in the fall of the year, selling them mostly to market, although he said he was having a considerable sale for these year-and-a-half old birds for layers; as he buys the nearly matured pullets at a per pound price and they increase in weight somewhat in the course of a year, he doubtless gets about as much as he pays for them. When asked if he averaged to make a dollar apiece profit, Mr. Hayward replied: "No, not quite so much as that," and in reply to questions about the health of the flocks that are kept in such close confinement, he said he never had a frosted comb, or any such trouble, that he lost now and then a bird (by death), perhaps ten to twelve per cent, but as that was a common experience to every one who kept fowls he did not let that trouble him. This statement as to losses was corroborated by two of the men on the place with whom we talked. They said it was the practice to dig a pit in the fall at some convenient place in the woods, throw the dead hens in it along through the winter and in the spring bury them with the earth that had been thrown out. Such a mortality looks to be a tremendous loss when we go into the big figures; of course it is a heavy loss, say three or four hundred dollars' worth of poultry meat, but the per cent of deaths is the same as if a man had two or three die out of his flock of two dozen.

DROPPINGS REMOVED TWICE A YEAR

The droppings are cleaned out of the houses twice a year, fall and spring; once a week (or thereabouts) a shovelful of dry earth is scattered over the droppings in each house and the piles left to accumulate till the next cleaning. There was here and there a flock with some evidence of feather eating, but as a whole the birds looked hearty and healthy; quite as much so as the average flocks one will see on farms where they have outside runs. It ought to be mentioned that two of the causes of the losses by death mentioned above were that sometimes three or four birds would pile on top of one that

85—WELL ARRANGED POULTRY YARDS

clung persistently to the nest and the one at the bottom would be smothered, and that now and then a bird would be pecked by its mates until the blood was started and then they would pitch upon it and peck it to death.

We asked some of the men if the egg-eating vice troubled them and could get no acknowledgment that it did; but as the houses are visited but twice a day and then but for a moment to each it is possible that quite a few eggs might disappear in that way without their knowledge, The fact that there was an egg lying on the floor in each of four of the houses we looked into and two eggs outside the nest boxes in one house would make it easy to suspect egg-eating. Their accidentally stepping upon and breaking an egg on the hard floor of the pen would cause that egg to be eaten instantly and the step from accidental breaking and eating to purposely breaking and eating is very easy.

As would be expected in such a case, the stock is mostly the common farm scrubs—or "dunghills," with a preponderance of Leghorn blood in it, but here and there in the houses we looked into, some fairish Barred Plymouth Rocks were to be seen; this fact is significant of the widespread popularity of that variety and bears out the claim of a well-known poultryman, that the Barred Rock is the most generally popular variety of fowls in America. If he had said "New England" he would probably have been right.

INDEX

Bone and Meat Meal; Green..........76
Breeding for Egg Production; Pedigree..........43
Breeding for Eggs..........43
Cold Storage; Preservation of Eggs in..........84
Demand and Supply..........24
Eastern Egg Market; The..........26
Egg-Eating Cured..........68
Egg Farm; A Large..........94
Egg Farm; A White Leghorn..........36
Egg Farms; Profitable..........87
Egg Farms; Rhode Island..........87
Egg Industry; The..........9
Egg Market; The Eastern..........26
Egg Market; The Western..........31
Egg Producers; Minorcas as..........49
Egg Production Influenced by Food..........73
Egg Production; Pedigree Breeding for..........43
Egg Production; Successful..........9
Egg Production; Winter..........67
Egg Ration; An..........81
Egg Yielding Capacity of Hens..........44
Eggs; A Practical Study of..........58
Eggs at Minimum Cost..........71
Eggs; Breeding for..........43
Eggs by Express; Shipping..........35
Eggs; Experiment in Feeding for..........79
Eggs; Experiments Layers and..........69
Eggs; Feeding for..........58
Eggs for Market; Poultry and..........92
Eggs for Profit..........36
Eggs in Cold Storage; Preservation of..........84
Eggs in Incubation; Weight of..........77
Eggs; Preservation of..........72
Eggs; Private Trade in..........38
Eggs; Remunerative Prices for..........40
Eggs; Table Scraps Into..........63
Eggs; The Quality of..........33
Eggs; Waste Products Into..........62
Experiments—Layers and Eggs..........69

Express; Shipping Eggs by..........35
Farmers's Opportunity; The..........24
Feeding for Eggs..........58
Feeding For Eggs; Experiments on..........69
Fertility; Heavy Laying and..........64
Feeding; Rations and..........60
Food; Egg Production Influenced By..........73
Heavy Laying and Fertility..........64
Introduction..........7
Laying and Fertility; Heavy..........64
Leghorn Egg Farm; A White..........36
Layers and Eggs; Experiments..........69
Layers and More of Them; Better..........51
Layers; Pullets vs Hens as..........78
Market; Poultry and Eggs for..........92
Meat Meal; Green Bone and..........76
Minorcas as Egg Producers..........49
Molt; Producing an Early..........77
Pedigree Breeding for Egg Production..........43
Poultry and Eggs for Market..........92
Preservation of Eggs..........82
Preservation of Eggs in Cold Storage..........84
Prices for Eggs; Remunerative..........40
Profitable Egg Farms..........87
Pullets vs Hens as Layers..........78
Quality of Eggs; The..........33
Rations and Feeding..........60
Remunerative Prices for Eggs..........40
Rhode Island Egg Farms..........87
Shipping Eggs by Express..........35
Successful Egg Production..........9
Table Scraps Into Eggs..........63
Trade in Eggs; Private..........38
Trap Nests a Necessity..........56
Waste Products Into Eggs..........62
Weight of Eggs in Incubation..........77
Western Egg Market; The..........31
Winter Egg Production..........67

www.ingramcontent.com/pod-product-compliance
Lightning Source LLC
Chambersburg PA
CBHW082343220526
45470CB00008B/2624